Collins

OCR GCSE Revision

Chemistry

Chemistry

OCR Gateway GCSE

Revision Guide

Eliot Attridge

Contents

HT **Higher Tier Content**

Contents

Paper 2   Predicting and Identifying Reactions and Products

HT Higher Tier Content

Contents

Contents

HT Higher Tier Content

Review Questions

Recap of KS3 Key Concepts

1 Write down whether each substance in the table is an **element**, **mixture** or **compound**.

Name of Substance	Type of Substance
Distilled water	
Gold	
Glucose	
Salt water	

[4]

2 **a)** Look at the following chemical reactions and tick (✓) the **two** reactions that are written correctly.

    A magnesium + sulfuric acid → magnesium nitrate + hydrogen

    B copper oxide + hydrochloric acid → copper chloride + water

    C iron + phosphoric acid → iron phosphate + hydrogen

    D lithium oxide + nitric acid → lithium nitrate + hydrogen

[2]

  b) When magnesium is burned in air, it produces a bright white light and a white, powdery residue.

    Write the **word equation** for the reaction.

[2]

3 Draw **three** diagrams to show the arrangement of particles in a solid, a liquid and a gas. [3]

4 For a metal to increase in temperature, its atoms must gain energy.

What type of energy must the metal atoms gain?

  A Nuclear energy   **C** Chemical energy

  B Heat energy   **D** Kinetic energy   [1]

5 Linda is stung by a bee.
The bee's sting is acidic.

  a) What chemical is the opposite of an acid?   [1]

  b) Which of the following could neutralise an acid?

    A Water   **C** Vinegar

    B Baking soda   **D** Magnesium   [1]

6 Gold and platinum are among a few metals that can be found in their elemental form on Earth.

    a) Explain why these metals are found in their elemental form. [2]

    b) Suggest why aluminium was discovered much later in human history than gold and platinum. [4]

7 Ruth is using universal indicator to identify acids and bases.

Suggest what colour universal indicator would turn for each of the examples below.

    a) Stomach acid [1]

    b) Juice of a lime [1]

    c) Tap water [1]

    d) Hand soap [1]

8 The diagram below shows the arrangement of atoms in four different substances.

A 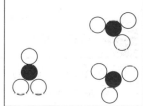   C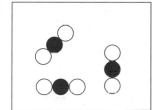

B    D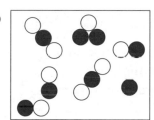

    a) Which box has a substance that is a mixture of compounds? [1]

    b) Which box has a substance that could be carbon dioxide? [1]

    c) Which box has a substance that could be a pure element? [1]

9 In winter fish are able to survive in a fish pond, even when the fish pond freezes over.

Explain what property of water enables the fish to survive. [2]

Total Marks _____ / 29

Particle Model and Atomic Structure

You must be able to:

- Use the particle model to explain states of matter and changes in state
- Describe the structure of the atom
- Explain how the atomic model has changed over time
- Calculate the number of electrons, protons and neutrons in atoms, ions and isotopes.

Particle Theory of Matter

- The particle theory of matter is based on the idea that all matter is made up of small particles.
- It is used to explain the structure and properties of solids, liquids and gases.

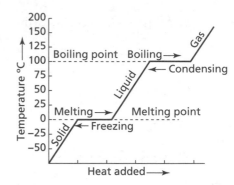

States of Matter

- When particles are heated (given energy), the energy causes them to move more.
- Even in a solid, the particles vibrate.
- At 0°C, water changes **state** from solid to liquid. This is called the melting point.
- The water **molecules** move faster as the water is heated.
- At 100°C, the water molecules change from liquid to gas. This is called the boiling point.
- When a substance is changing state, the temperature remains constant.
- If gaseous water is cooled, the molecules slow down as they lose kinetic energy and liquid water forms.
- Changes in state are physical changes – it is relatively easy to go back to a previous state.
- Chemical changes are permanent changes where an atom or molecule chemically joins to another atom or molecule.
- The particle model uses circles or spheres to represent particles.

> HT The particle model has some limitations.
>
> HT It fails to address the forces of attraction between particles, the size of the particles and that there is space between the particles.

The Atom

- The **atom** is the smallest part of an element that retains the properties of that element.
- Atoms are very small, with a radius of about 1^{-10}m and a mass of about 10^{-23}g.
- An atom has a central, positively charged nucleus, surrounded by shells of negatively charged electrons.
- The radius (width) of the nucleus is far smaller than that of the whole atom.
- The nucleus is made up of protons and neutrons.
- An atom has no overall charge because the charges of the (positive) protons and (negative) electrons cancel each other out.

> **Key Point**
>
> The particles within a substance only have no movement at absolute zero (−273.15°C). At all other temperatures the particles will vibrate.

Atomic Particle	Relative Charge	Relative Mass
Proton	+1	1
Neutron	0	1
Electron	−1	0.0005 (zero)

> **Key Point**
>
> No matter how many electrons or neutrons there are in an atom, it is the number of protons that defines an element and its properties.

The History of the Atom

- The model of atomic structure has changed over time.
- In the early 1800s, John Dalton proposed that all atoms of the same element were identical.
- In 1897, J. J. Thomson discovered the electron and put forward the 'plum pudding model' with charges spread evenly throughout the atom.
- In 1911, Ernest Rutherford with his assistants, Geiger and Marsden, discovered that the atom had a dense centre – the nucleus – by firing **alpha particles** at gold foil.
- In 1913, Niels Bohr predicted that electrons occupy energy levels or orbitals.

Mass Number and Atomic Number

- The **mass number** is the total number of protons and neutrons in an atom.
- The **atomic number** is the number of protons in an atom.
- An **ion** is formed when an element gains or loses electrons.

Element	Symbol	Mass Number	Atomic Number	Protons	Neutrons	Electrons
Hydrogen	$^{1}_{1}H$	1	1	1	0	1
Helium	$^{4}_{2}He$	4	2	2	2	2
Sodium	$^{23}_{11}Na$	23	11	11	12	11
HT Oxide	$^{16}_{8}O^{2-}$	16	8	8	16 – 8 = 8	8 + 2 = 10

number of protons =
        atomic number
number of electrons =
        number of protons (in an atom)
number of neutrons =
        mass number – atomic number

The oxide ion is formed when an oxygen atom gains two extra electrons.

- **Isotopes** are atoms of the same element that have:
  - the same atomic number
  - a different mass number, due to a different number of neutrons.
- For example, carbon has three main isotopes:

Isotope	Symbol	Mass Number	Atomic Number	Protons	Neutrons	Electrons
Carbon-12	$^{12}_{6}C$	12	6	6	6	6
Carbon-13	$^{13}_{6}C$	13	6	6	7	6
Carbon-14	$^{14}_{6}C$	14	6	6	8	6

Taking away the atomic number from the mass number gives the number of neutrons. So, there are two extra neutrons in chlorine-37 compared to chlorine-35.

Chlorine has two isotopes:

$^{35}_{17}Cl$	Mass number = 35
	Atomic number = 17

$^{37}_{17}Cl$	Mass number = 37
	Atomic number = 17

35 – 17 = 18 neutrons   37 – 17 = 20 neutrons

> ### Quick Test
>
> 1. What are isotopes?
> 2. How many neutrons does an element have with a mass number of 14 and an atomic number of 6?
> 3. Why have different models of the atom been proposed over time?

Purity and Separating Mixtures

You must be able to:

- Suggest appropriate methods to separate substances
- Work out empirical formulae using relative molecular masses and relative formula masses
- Calculate the R_f values of different substances that have been separated using chromatography.

Purity

- In chemistry something is pure if all of the particles that make up that substance are the same, e.g. pure gold only contains gold atoms and pure water only contains water molecules.
- All substances have a specific melting point at room temperature and pressure.
- Comparing the actual melting point to this known value is a way of checking the purity of a substance.
- Any impurities cause the substance to melt at a different temperature.
- **Formulations** are mixtures that have been carefully designed to have specific properties, e.g. alloys (see page 81).

> ### Key Point
>
> In the world outside the lab, 'pure' is often used to describe mixtures, e.g. milk. This means that nothing has been added; it does not indicate how *chemically* pure it is.

Relative Atomic, Formula and Molecular Mass

- Every element has its own **relative atomic mass (A_r)**.
- This is the ratio of the average mass of one atom of the element to one-twelfth of the mass of an atom of carbon-12.
- The **relative molecular mass (M_r)** is the sum of the relative atomic masses of each atom making up a molecule.
- For example, the M_r of O_2 is $2 \times 16 = 32$.
- The **relative formula mass (M_r)** is the sum of the relative atomic masses of all the atoms that make up a compound.

For example, the relative atomic mass of magnesium is 24 and of oxygen is 16.

Calculate the relative formula mass of H_2O.

H: $2 \times 1 = 2$

O: $1 \times 16 = 16$

H_2O: $2 + 16 = 18$

> Multiply the number of atoms of each element in the molecule by the relative atomic mass.

> Add them all up to calculate the M_r.

Empirical Formula

- The empirical formula is the simplest whole number ratio of each type of atom in a compound.
- It can be calculated from the numbers of atoms present or by converting the mass of the element or compound.

> ### Key Point
>
> Always show your working-out when calculating empirical formulae. You will be less likely to make mistakes if you do.

What is the empirical formula of a compound with the formula $C_6H_{12}O_6$?

$$C = \frac{6}{6} = 1 \qquad H = \frac{12}{6} = 2 \qquad O = \frac{6}{6} = 1$$

The empirical formula is written as CH_2O.

> Work out the smallest ratio of whole numbers by dividing each by the smallest number. This would be $C_1H_2O_1$.

> Remember, the 1 is not written.

- For example, all alkenes (see page 84) have the empirical formula C_1H_2 although the 1 is not written, so it would appear as CH_2.

What is the empirical formula of a compound containing 24g of carbon, 8g of hydrogen and 32g of oxygen?

Elements	Carbon	:	Hydrogen	:	Oxygen
Mass of element	24	:	8	:	32
A_r of element	12	:	1	:	16
$\dfrac{\text{Mass of element}}{A_r}$	2	:	8	:	2
Divide by smallest number	÷ 2		÷ 2		÷ 2
Ratio of atoms in empirical formula	1	:	4	:	1

List all the elements in the compound.

To find the number of moles, divide the mass of each element by its relative atomic mass.

Divide each answer by the smallest number in step 2 to obtain a ratio.

The ratio may have to be scaled up to give whole numbers.

The empirical formula is therefore CH_4O.

Remember, it is incorrect to write the 1 for an element.

Separation Techniques

- Techniques that can be used to separate mixtures include:
- **Filtration** – a solid is separated from a liquid (e.g. copper oxide solid in copper sulfate solution).
- **Crystallisation** – a solvent is evaporated off to leave behind a solute in crystal form (e.g. salt in water).
- **Distillation** – two liquids with significantly different boiling points are separated, i.e. when heated, the liquid with the lowest boiling point evaporates first and the vapour is condensed and collected.
- **Fractional distillation** – a mixture of liquids with different boiling points are separated (e.g. petrol from crude oil).
- **Chromatography** – substances in a mixture are separated using a **stationary phase** and a **mobile phase**.
  - **Paper chromatography** – this is useful for separating mixtures of dyes in solution (e.g. dyes in ink).
  - **Thin layer chromatography (TLC)** – this is more accurate than paper chromatography and uses a thin layer of an inert solid for the stationary phase.
  - **Gas chromatography** – this separates gas mixtures by passing them through a solid stationary phase.
- Substances separated by chromatography can be identified by calculating their R_f **values**.

$$R_f = \frac{\text{distance moved by the compound}}{\text{distance moved by the solvent}}$$

- Separated substances can be identified by comparing the results to known R_f values.

> **Key Point**
>
> Substances move up the stationary phase at different rates depending upon their properties. The rate will remain the same as long as the conditions are the same.

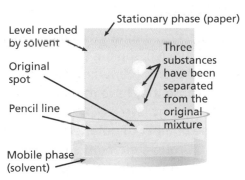

Stationary phase (paper)

Level reached by solvent

Original spot

Pencil line

Mobile phase (solvent)

Three substances have been separated from the original mixture

> **Key Words**
>
> formulations
> relative atomic mass (A_r)
> relative molecular mass (M_r)
> relative formula mass (M_r)
> chromatography
> stationary phase
> mobile phase
> R_f value

Quick Test

1. What is the relative formula mass of $Mg(OH)_2$?
2. What is paper chromatography used to separate?
3. What is the empirical formula of a compound with the formula C_2H_6?

Bonding

You must be able to:

- Explain how metals and non-metals are positioned in the periodic table
- Describe the electronic structure of an atom
- Draw dot and cross diagrams for ions and simple covalent molecules.

The Periodic Table

- An element contains one type of atom.
- Elements cannot be chemically broken down into simpler substances.
- There are about 100 naturally occurring elements.
- The design of the modern periodic table was first developed by Mendeleev.
- Elements in Mendeleev's table were placed into groups based on their **atomic mass**.
- Mendeleev's method was testable and predicted elements not yet discovered.
- However, some elements were put in the wrong place because the values used for their atomic masses were incorrect.
- The modern periodic table is a modified version of Mendeleev's table.
- It takes into account the arrangement of electrons, the number of electrons in the outermost shell, and atomic number.

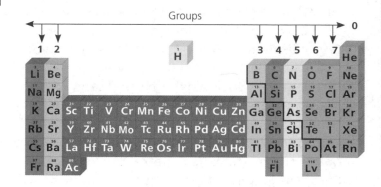

Groups

- A vertical column of elements in the periodic table is a **group**.
- Lithium (Li), sodium (Na) and potassium (K) are in Group 1.
- Elements in the same group have similar chemical properties because they have the same number of electrons in their **outer shell** (or energy level).
- The number of outer electrons is the same as the group number:
  - Group 1 elements have one electron in their outer shell.
  - Group 7 elements have seven electrons in their outer shell.
  - Group 0 elements have a full outer shell.

Periods

- A horizontal row of elements in the periodic table is a **period**.
- Lithium (Li), carbon (C) and neon (Ne) are in Period 2.
- The period for an element is related to the number of occupied electron shells it has.
- For example, sodium (Na), aluminium (Al) and chlorine (Cl) have three shells of electrons so they are in the Period 3.

Key Point

The number of protons in a nucleus of an element never changes. That's why the periodic table shows the atomic number.

Metals and Non-Metals

- The majority of the elements in the periodic table are metals.
- Metals are very useful materials because of their properties:
  - They are lustrous, e.g. gold is used in jewellery.
  - They are hard and have a high density, e.g. titanium is used to make steel for drill parts.
  - They have high tensile strength (are able to bear loads), e.g. steel is used to make bridge girders.
  - They have high melting and boiling points, e.g. tungsten is used to make light-bulb filaments.
  - They are good conductors of heat and electricity, e.g. copper is used to make pans and wiring.
- Metals can react with non-metals to form ionic compounds.
- For example, metals react with oxygen to form metal oxides.

Electronic Structure

- An element's position in the periodic table can be worked out from its **electronic structure**.
- For example, sodium's electronic structure is 2.8.1 (atomic number = 11):
  - It has three orbital shells, so it can be found in Period 3.
  - It has one electron in its outer shell, so it can be found in Group 1.
- The electronic structure can also be shown using a dot and cross diagram, in which each cross represents an electron.

Sodium atom, Na
2.8.1

Sodium ion, Na$^+$
[2.8]$^+$

+ 1e$^-$

Chemical Bonds

- Chemical bonds are *not* physical structures.
- They are the transfer or sharing of electrons, which leads to the atoms involved becoming more stable.
- An **ionic bond** is formed when one or more electrons are donated by one atom or molecule and received by another atom or molecule.
- When an ionic compound is in solution, or in a molten state, the ions move freely.
- When an ionic compound is solid, ions are arranged in a way to cancel out the charges.
- A **covalent bond** is formed when atoms share electrons to complete their outermost shell.

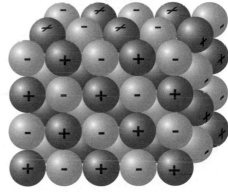

 Positively charged ion ⊙ Negatively charged ion

Quick Test

1. Give **two** ways that elements are arranged in the modern periodic table.
2. Draw a dot and cross diagram to show the electronic structure of magnesium.
3. Write the electronic structure for chlorine.

Key Words

atomic mass
group
outer shell
period
electronic structure
ionic bond
covalent bond

Models of Bonding

You must be able to:

- Describe and compare the type of bonds in different substances and their arrangement
- Use a variety of models to represent molecules
- Identify the limitations of different models.

Models of Bonding

- **Models** can be used to show how atoms are bonded together.
- Dot and cross diagrams can show:
  - each shell of electrons or just the outer shell
  - how electrons are donated or shared.
- Methane is a **covalent** compound. Each molecule is made up of a carbon atom joined to four hydrogens (CH_4).

Methane, CH_4

Methane, CH_4

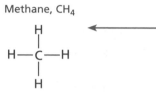

> Each line or shared pair of electrons shows a covalent bond.

Methane, CH_4

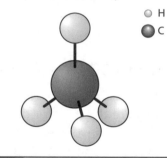

○ H
● C

- Ball and stick models give an idea of the 3D shape of a molecule or compound.
- Each model has limitations:
  - The scale of the nucleus to the electrons is wrong in most models.
  - Models show bonds as physical structures.
  - Most models do not give an accurate idea of the 3D shape of a molecule.
  - The bond lengths are not in proportion to the size of the atoms.
  - Models aid our understanding about molecules, but they are not the real thing.

> **Key Point**
>
> Scientists use models to help solve problems. As atoms are too small to be seen with the naked eye, models are a helpful way of visualising them.

Ion Formation

- Metals give away electrons to become positive ions:

Sodium atom, Na
2.8.1

Sodium ion, Na⁺
[2.8]⁺

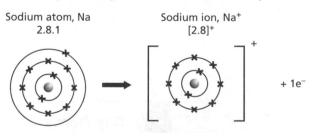

+ 1e⁻

> Sodium gives away a single electron to become a Na⁺ ion.

- Non-metals gain electrons to become negative ions:

Chlorine atom, Cl
2.8.7

Chloride ion, Cl⁻
[2.8.8]⁻

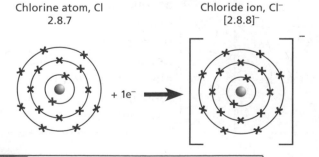

+ 1e⁻

> Chlorine gains an electron to become a Cl⁻ ion.

- **Ionic bonds** are the electrostatic forces of attraction that hold the ions together.

Simple Molecules

- When non-metals or non-ionic molecules join together, the atoms share electrons and form a covalent bond. These are called **simple molecules**.
- Hydrogen gas, H_2, is a covalent molecule.

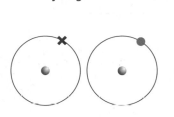
Hydrogen Atoms

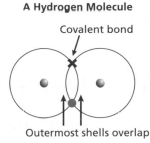
A Hydrogen Molecule
Covalent bond
Outermost shells overlap

Giant Covalent Structures

- **Giant covalent structures** are formed when the atoms of a substance form repeated covalent bonds.
- Silicon dioxide is a compound made up of repeating silicon and oxygen atoms joined by single covalent bonds.

Polymers

- A **polymer** (see page 87) is formed when repeated units are covalently bonded together.
- For example, when lots of ethene molecules are joined together they form poly(ethene).

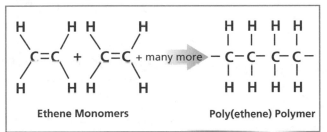
Ethene Monomers   Poly(ethene) Polymer

Metals

- Metal atoms are held together by strong metallic bonds.
- The metal atoms lose their outermost electrons and become positively charged.
- The electrons can move freely from one metal ion to another.
- This causes a sea of **delocalised** (free) electrons to be formed.

Key Point

The term 'ionic bond' suggests there is a permanent, physical link between ions. However, when in solution or molten, the ions move further away from each other.

Each hydrogen atom now has a full outermost shell, with two electrons.

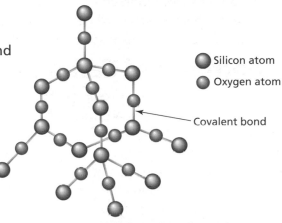
○ Silicon atom
○ Oxygen atom
Covalent bond

Atoms are shown by their element symbol. Bonds are shown with lines. Two lines together indicate a double bond (two covalent bonds between atoms).

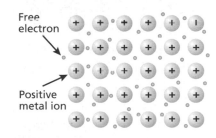
Free electron

Positive metal ion

Key Words

model
covalent
ionic bond
simple molecule
giant covalent structure
polymer
delocalised

Quick Test

1. What is meant by the term 'sea of delocalised electrons'?
2. Give **two** limitations of a dot and cross model of a covalent compound.
3. What is meant by the term 'giant covalent structure'?

Properties of Materials

You must be able to:

- Describe how carbon can form a wide variety of different molecules
- Explain the properties of diamond, graphite, fullerenes and graphene
- Explain what nanoparticles are and the risks they pose.

Carbon

- Carbon is the sixth element in the periodic table and has an atomic mass of 12.
- Carbon is in Group 4 because it has four electrons in its outer shell.
- This means that it can make up to four covalent bonds with other atoms.
- It can also form long chains of atoms and rings.
- There is a vast variety of naturally occurring and synthetic (man-made) carbon-based compounds, called **organic** compounds.

Allotropes of Carbon

- Each carbon atom can bond with up to four other carbon atoms.
- Different structures are formed depending on how many carbon atoms bond together.
- These different forms are called **allotropes** of carbon. They do not contain any other elements.
- Graphite is formed when each carbon atom bonds with three other carbon atoms:
  - Graphite has free electrons so it can conduct electricity, e.g. in electrolysis.
  - The layers are held together by weak bonds, so they can break off easily, e.g. in drawing pencils and as a dry lubricant.
- **Graphene** is a single layer of graphite:
  - In this form, the carbon is 207 times stronger than steel.
  - Graphene has free electrons so it can conduct electricity.
  - It is used in electronics and solar panels.
- Diamond is formed when each carbon atom bonds with four other carbon atoms:
  - Diamond cannot conduct electricity, as all its outermost electrons are involved in bonding.
  - Diamonds are very hard. They are used in drill bits and polished diamonds are used in jewellery.
  - Diamond is extremely strong because each atom forms the full number of covalent bonds.

Relative atomic mass

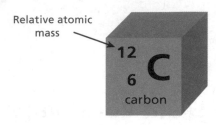

Key Point

There are a few carbon compounds that are non-organic. They include the oxides of carbon, cyanides, carbonates and carbides.

Graphite

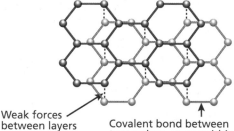
Weak forces between layers
Covalent bond between two carbon atoms within a layer

Graphene

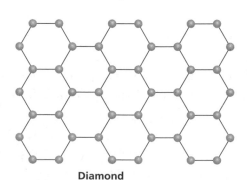

Diamond

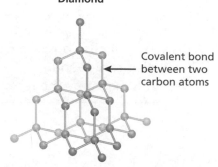
Covalent bond between two carbon atoms

- **Fullerenes** are tubes and spherical structures formed using only carbon atoms:
  - They are used as superconductors, for reinforcing carbon-fibre structures, and as containers for drugs being introduced into the body.

Bonding and Changing State

- Bonding is an attraction between atoms in elements and compounds.
- If the attraction is weak, then it is easy to separate the atoms compared to those with a stronger attraction.
- The ions in ionic substances are more easily separated when they are in solution or molten, as they can move about freely.
- When an ionic substance is in its crystal (solid) form, i.e. when the distance between ions is at its smallest, it is very difficult to separate the ions due to the strong electrostatic forces.
- They form a giant lattice structure.
- The melting point of ionic substances is, therefore, very high.
- For example, the melting point of NaCl is 801°C.
- Covalent bonds are very strong.
- If there are a lot of bonds, e.g. in a giant covalent compound, the melting point will be very high (higher than for ionic compounds).
- For example, graphite melts at 3600°C.
- Simple covalent molecules have very low boiling points.
- The simplest gas, hydrogen, melts at –259°C and has a boiling point of –252.87°C.
- This is because the **intermolecular forces** that hold all the molecules together are weak and, therefore, easily broken.

Nanoparticles

- **Nanoparticles** are particles with a size between 1 and 100nm.
- Hydrogen atoms, by comparison are 0.1nm wide.
- At this size range all materials lose their **bulk properties**.
- For example, copper is bendy above 50nm, but nanoparticles of copper are ultra-strong and cannot be bent.
- Using nanoparticle materials opens up a new range of properties.
- For this reason, they are increasingly being used in a wide variety of industries, from medicine to construction.
- Nanoparticles exist naturally and can also be manufactured.
- They are small enough to enter respiratory systems and could potentially cause damage.
- They have a very high surface area compared to their volume, so they can act as catalysts.
- Silver nanoparticles can kill bacteria, both good and bad. The effect on the immune system is not known.

Structure of Buckminsterfullerene

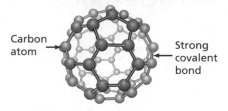

Carbon atom → ← Strong covalent bond

> **Key Point**
>
> Don't confuse intermolecular forces (the forces between molecules) with the intramolecular forces (e.g. the covalent bonds between the atoms in the molecules).

> **Key Point**
>
> There are 1 million nm per mm, 10 million nm per cm and 1 billion nm per m.

> **Key Point**
>
> It is important that scientists consider the risks and benefits of new technologies before introducing them to the outside world.

> **Key Words**
>
> organic
> allotropes
> graphene
> fullerenes
> intermolecular force
> nanoparticle
> bulk properties

> **Quick Test**
>
> 1. What is meant by the term 'organic compound'?
> 2. Why is diamond so strong?
> 3. What is the size range for a nanoparticle?

Practice Questions

Particle Model and Atomic Structure

1 Look at the heating curve for water.

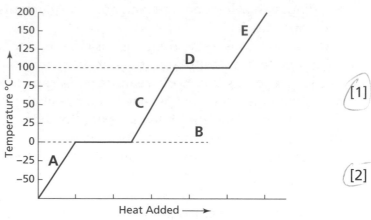

a) Which point, **A, B, C, D** or **E**, indicates the boiling point? D ✓ [1]

b) Describe what will happen to the particles in water as the temperature changes from −50°C to −25°C. [2]

The particles will start to vibrate more as they have more energy to do so

2 **HT** Look at the diagram of water molecules in distilled water.

What is in the gaps between the water molecules in the diagram? [1]

Nothing ✓

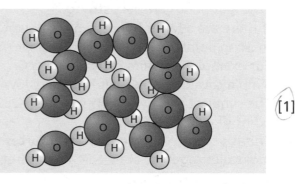

3 Look at the following elements on your periodic table and answer the questions below: **lithium, oxygen, neon, silicon** and **calcium**.

a) Which element has a mass number of 20? Neon ✓ [1]

b) Which element has an atomic number of 20? Calcium ✓ [1]

c) What would the mass number of an isotope of Si be if it had two extra neutrons? 30 ✓ [1]
(28 + 2)

4 Water vapour at 140°C was cooled to −40°C.

a) Draw a labelled diagram showing the changes that take place during the cooling process. [4]

b) Describe what happens to the particles in the water as it cools. [2]
They would vibrate less as they have less energy and may get closer and form a more regular pattern

5 What are the charges and relative masses of: [6]

a) An electron? b) A proton? c) A neutron?

   −1   +1   0
   0.0005   1   1

Total Marks _____19_____ / 19

Purity and Separating Mixtures

1 a) What does the term **pure** mean in chemistry? *It is made up of only one type of atom.* [1]

   b) Describe how melting points can be used to help identify a pure substance. *If the substance does not melt at the expected temp* ✓ [2]

2 Athina is separating food colourings using chromatography.

   a) Calculate the R_f value for the two colours in **X**. Show your working. *7.5/28 = 0.27* ✓ *18/28 = 0.64* [3]

   b) Which of the food colourings, **A, B, C, D** or **E**, matches **X**? *D* [1]

(chromatography diagram, scale 0–30, Solvent front, Distance travelled by solvent, Spot origin line ('start line'), lanes X A B C D E)

3 A compound has the formula $C_8H_8S_2$.

   What is its empirical formula? C_4H_4S [1]

4 What is the empirical formula of a compound containing 84g of carbon, 16g of hydrogen and 64g of oxygen? Show your working. $C_{21}H_4O_{16}$ $C_7H_{16}O_4$ [3]

5 What is the best method for separating water from an ink solution?

   Ⓐ Distillation   **C** Evaporation

   B Filtration   **D** Chromatography [1]

Total Marks _____ 9 _____ / 12

Bonding

1 Look at the following chemical symbols from the periodic table.

   a) Write down the atomic number for each element. [2]

   b) Potassium oxide has the formula: K_2O.
      Work out the **relative formula mass** for K_2O. *39.1 × 2 = 78.2 +16 = 94.2* [1]

   c) Calculate the **relative molecular mass** for oxygen. *16 × 2 = 32* [1]

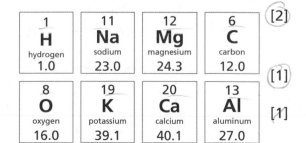

1	11	12	6
H	**Na**	**Mg**	**C**
hydrogen	sodium	magnesium	carbon
1.0	23.0	24.3	12.0
8	19	20	13
O	**K**	**Ca**	**Al**
oxygen	potassium	calcium	aluminum
16.0	39.1	40.1	27.0

Practice Questions

2. a) Why do elements in the same group of the periodic table have the same properties? [1]

    They have the same no. valence electrons.

    b) Lithium, carbon and neon are in the same period.

    What feature do these elements share? *They have two shells* [1]

3. The electronic structure of potassium, K, is written as 2.8.8.1.

    a) Which of the following dot and cross diagrams represents K?

    A   B   C   D   E   F

    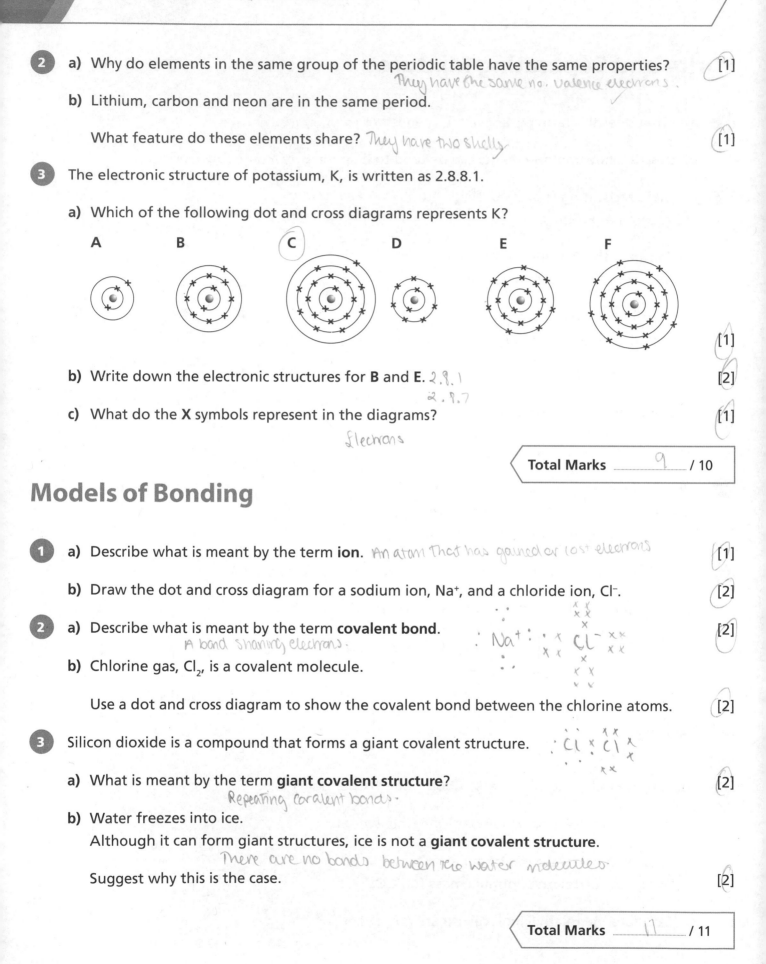

    [1]

    b) Write down the electronic structures for **B** and **E**. *2.8.1* [2]

    2.8.7

    c) What do the **X** symbols represent in the diagrams? [1]

    Electrons

    > Total Marks _____ 9 / 10

Models of Bonding

1. a) Describe what is meant by the term **ion**. *An atom that has gained or lost electrons* [1]

    b) Draw the dot and cross diagram for a sodium ion, Na⁺, and a chloride ion, Cl⁻. [2]

2. a) Describe what is meant by the term **covalent bond**. [2]

    A bond sharing electrons.

    b) Chlorine gas, Cl_2, is a covalent molecule.

    Use a dot and cross diagram to show the covalent bond between the chlorine atoms. [2]

3. Silicon dioxide is a compound that forms a giant covalent structure.

    a) What is meant by the term **giant covalent structure**? [2]

    Repeating covalent bonds.

    b) Water freezes into ice.

    Although it can form giant structures, ice is not a **giant covalent structure**.

    Suggest why this is the case.

    There are no bonds between the water molecules. [2]

    > Total Marks _____ 11 / 11

Properties of Materials

1 **a)** Explain how carbon can form a variety of different molecules. *It has 4 valence electrons* [3]

b) Describe what is meant by the term **allotrope**. *A different forms of an element* [1]

c) Give the names of **three** allotropes of carbon. *Graphite, diamond, fullerene* [3]

2 Graphene is used in electronics and solar panels.

Graphene

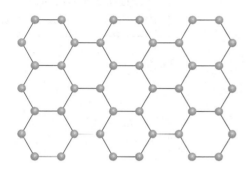

a) Explain why graphene is used for these purposes. *It conducts electricity* [2]

b) Other than cost, explain why diamond is **not** used for these purposes. *It does not conduct electricity* [1]

c) Draw the structure of diamond. [2]

3 Ionic compounds can conduct electricity.

a) Describe the conditions required for an ionic compound to conduct electricity. *Molten or dissolved* [2]

b) Why do ionic compounds in their crystalline form typically have very high melting points? [2]

4 Simple covalent compounds typically have very low melting points.

Which of the following would have the lowest melting point?

A CH_4

B H_2O

C H_2

D O_2 [1]

Total Marks ___13___ / 17

Introducing Chemical Reactions

You must be able to:

- Use names and symbols to write formulae and balanced chemical equations
- Describe the states of reactants and products in a chemical reaction.

Law of Conservation of Mass

- The law of conservation of mass means that no atoms are created or destroyed.
- This means that, in a chemical reaction, the mass of the **products** will always equal the mass of the **reactants**.
- The atoms in a reaction can recombine with other atoms, but there will be no change in the overall number of atoms.
- This allows chemists to make predictions about chemical reactions. For example:
  - What might be formed when chemicals react together?
  - How much of the chemical or chemicals will be made?

Key Point

Chemicals are not 'used up' in a reaction. The atoms are rearranged into different chemicals.

Formulae and State Symbols

- Compounds can be represented using **formulae**, which use symbols and numbers to show:
  - the different elements in the compound
  - the number of atoms of each element in a molecule of the compound.
- A small subscript number following a symbol is a multiplier – it tells you how many of those atoms are present in a molecule.
- If there are brackets around part of the formula, everything inside the brackets is multiplied by the number on the outside.

> Sulfuric acid has the formula H_2SO_4.
> This means that there are two hydrogen atoms, one sulfur atom and four oxygen atoms.

The ratio of the number of atoms of each element in sulfuric acid is 2H : 1S : 4O.

> $Ca(NO_3)_2$
> This means that there is one calcium atom and two nitrate (NO_3) groups.
> In total there are one calcium, two nitrogen and six oxygen atoms present in this compound.

- There are four state symbols, which are written in brackets after the formula symbols and numbers:
  - (s) = **solid**
  - (l) = **liquid**
  - (g) = **gas**
  - (aq) = **aqueous** (dissolved in water).

$$H_2O(l) \qquad CO_2(g) \qquad H_2SO_4(aq) \qquad S_8(s)$$

Balancing Equations

- Equations show what happens during a chemical reaction.
- The reactants are on the left-hand side of the equation and the products are on the right.
- Remember, no atoms are lost or gained during a chemical reaction so the equation must be balanced.
- There must always be the same number of each type of atom on both sides of the equation.
- A large number written before a molecule is a **coefficient** – it is a multiplier that tells you how many copies of that whole molecule there are.

$2H_2SO_4(aq)$ means there are two molecules of $H_2SO_4(aq)$ present.

- To balance an equation:

Reactants	→	Products	
magnesium + oxygen	→	magnesium oxide	← Write the word equation.
Mg + O_2	→	MgO	← Write the formulae of the reactants and products.
Mg + O O	→	Mg O	
Mg + O O	→	Mg O Mg O	← Balance the equation.
Mg Mg + O O	→	Mg O Mg O	
$2Mg(s)$ + $O_2(g)$	→	$2MgO(s)$	← Add state symbols.

- You should be able to balance equations by looking at the formulae without drawing the atoms. For example:

calcium carbonate	+	nitric acid	→	calcium nitrate	+	carbon dioxide	+	water
$CaCO_3$	+	HNO_3	→	$Ca(NO_3)_2$	+	CO_2	+	H_2O
$CaCO_3$	+	$2HNO_3$	→	$Ca(NO_3)_2$	+	CO_2	+	H_2O
$CaCO_3(s)$	+	$2HNO_3(aq)$	→	$Ca(NO_3)_2(aq)$	+	$CO_2(g)$	+	$H_2O(l)$

- Equations can also be written using displayed formulae (see page 84). These must be balanced too.

Key Point

If you find the numbers keep on increasing on both sides of an equation you are trying to balance, it is likely you have made a mistake. Restart by checking the formulae and then rebalancing the equation.

Key Words

products
reactants
formulae
solid
liquid
gas
aqueous
coefficient

Quick Test

1. What is the formula of calcium hydroxide?
2. Write the balanced symbol equation for the reaction: sodium + chlorine → sodium chloride.
3. How many of each atom are present in this formula: $2MgSO_4$?

Chemical Equations

You must be able to:

- Recall the formulae of common ions and use them to deduce the formula of a compound
- HT Use names and symbols to write balanced half equations
- HT Construct balanced ionic equations.

Formulae of Common Ions

- Positive ions are called **cations**. Negative ions are called **anions**.
- There are a number of common ions that have a set **charge**.
- See page 141 for a table of formulae of common ions.
- The roman numerals after a transition metal's name tell you its charge, e.g. iron(II) will have the charge Fe^{2+}.
- When combining ions to make an ionic compound, it is important that the charges cancel each other out so the overall charge is neutral.

> $Cu^{2+} + Cl^-$
>
> The formula is: $CuCl_2$

Key Point

Although ionic compounds are written as a formula (e.g. $CuCl_2$), they are actually dissociated when in solution, i.e. the ions separate from each other.

Two negative charges are needed to cancel the charge on the copper cation. These will come from having two chloride ions.

HT Stoichiometry

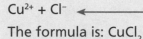

- **Stoichiometry** is the measurement of the relative amounts of reactants and products in chemical reactions.
- It is based on the conservation of mass, so knowing quantities or masses on one side of an equation enables you to work out the quantities or masses on the other side of the equation.

- For example, when magnesium is heated in air:
  $2Mg(s) + O_2(g) \rightarrow 2MgO(s)$
  - The mass of magnesium oxide formed is equal to the starting mass of magnesium plus the mass of oxygen from the air that is added to it.
- For example, when calcium carbonate is heated in air it thermally decomposes to form calcium oxide and carbon dioxide:
  $CaCO_3(s) \rightarrow CaO(s) + CO_2(g)$
  - The mass of calcium oxide remaining plus the mass of carbon dioxide added to the atmosphere is equal to the starting mass of calcium carbonate.

> HT When looking at the stoichiometry of a chemical reaction it is common to look at the ratios of the molecules and compounds to each other.

- The numbers needed to balance an equation can be calculated from the masses of the reactants and the products using moles.

> In a chemical reaction, 72g of magnesium was reacted with exactly 48g of oxygen molecules to produce 120g of magnesium oxide.
>
> Use the number of moles of reactants and products to write a balanced equation for the reaction.

$$\text{amount of Mg} = \frac{72}{24} = 3\text{mol}$$

$$\text{amount of O}_2 = \frac{48}{32} = 1.5\text{mol}$$

$$\text{amount of MgO} = \frac{120}{40} = 3\text{mol}$$

$$3Mg + 1.5O_2 \rightarrow 3MgO$$

$$2Mg + O_2 \rightarrow 2MgO$$

Use the masses of the reactants to calculate the number of moles present.

Divide the number of moles of each substance by the smallest number (1.5) to give the simplest whole number ratio.

This shows that 2 moles of magnesium react with 1 mole of oxygen molecules to produce 2 moles of magnesium oxide.

Limiting Reactants

- Sometimes when two chemicals react together, one chemical is completely used up during the reaction.
- When one chemical is used up, it stops the reaction going any further. It is called the **limiting reactant**.
- The other chemical, which is not used up, is said to be in excess.

Hydrogen ions to hydrogen gas:
1. Write formulae: $H^+ \rightarrow H_2$
2. Balance numbers: $2H^+ \rightarrow H_2$
3. Identify charges: 2^+ $\quad$ 0
4. Add electrons: $2H^+ + 2e^- \rightarrow H_2$

HT Half Equations

- **Half equations** can be written to show the changes that occur to the individual ions in a reaction:
  1. Write the formulae of the reactants and the products.
  2. Balance the number of atoms.
  3. Add the charges present.
  4. Add electrons (e^-) so that the charges on each side balance.

Chloride ions to chlorine gas:
1. $Cl^- \rightarrow Cl_2$
2. $2Cl^- \rightarrow Cl_2$
3. 2^- $\quad$ 0
4. $2Cl^- \rightarrow Cl_2 + 2e^-$

HT Balanced Ionic Equations

- When writing a balanced ionic equation, only the **species** that actually change form, i.e. gain or lose electrons, are written.
- The species that stay the same, the **spectator ions**, are ignored.
  1. Write the full balanced equation with state symbols.
  2. Write out all the soluble ionic compounds as separate ions.
  3. Delete everything that appears on both sides of the equation (the spectator ions) to leave the **net ionic equation**.

lead nitrate + potassium chloride ⟶

lead chloride + potassium nitrate

1. $Pb(NO_3)_2(aq) + 2KCl(aq) \longrightarrow PbCl_2(s) + 2KNO_3(aq)$
2. $Pb^{2+}(aq) + 2NO_3^-(aq) + 2K^+(aq) + 2Cl^-(aq) \longrightarrow$
   $PbCl_2(s) + 2K^+(aq) + 2NO_3^-(aq)$
3. $Pb^{2+}(aq) + 2Cl^-(aq) \longrightarrow PbCl_2(s)$

> **Key Point**
>
> It is convention to show added electrons only; the **electrons** being taken away are not shown.

The spectator ions, $NO_3^-(aq)$ and $K^+(aq)$, are removed.

This is the net ionic equation.

> **Key Words**
>
> cations
> anions
> charge
> limiting reactant
> HT stoichiometry
> HT half equation
> HT species
> HT spectator ions
> HT net ionic equation

> **Quick Test**
>
> 1. What are the formulae of barium oxide, copper fluoride and aluminium chloride?
> 2. Aluminium ions have a charge of 3^+ and oxide ions have a charge of 2^-. What is the formula of aluminium oxide?
> 3. HT What is the net ionic equation for the reaction of $Na_2CO_3(aq) + BaCl_2(aq)$?

Moles and Mass

You must be able to:

 Explain what a mole is

 Calculate the relative molecular mass, mass and number of moles of substances from equations and experimental results.

Moles

- In chemistry it is important to accurately measure how much of a chemical is present.
- Atoms are very small and there would be too many to count in even 1g of substance.
- Instead a measurement is used that represents a known, precise number of atoms – a **mole**.
- A mole represents a set amount of substance – the amount of substance that contains the same number of atoms as 12g of the element **carbon-12**.
- The number of atoms in 1 mole of carbon-12 is a very large number: 6.022×10^{23} atoms.
- This number is known as **Avogadro's constant**.

Calculations Using Moles

- Every element in the periodic table has an atomic mass.
- This means that the mass of one mole of an element will be equivalent to that element's **relative atomic mass** in grams (g).
- The mass of one mole of any compound is its relative formula mass (M_r) in g.
- The **relative molecular mass** of a compound is numerically the same as the relative formula mass. Its units are g/mol.
- You can use the following formulae to calculate the number of moles of an element or compound:

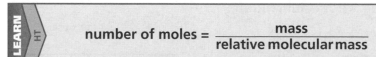

LEARN HT
$$\text{number of moles} = \frac{\text{mass}}{\text{relative molecular mass}}$$

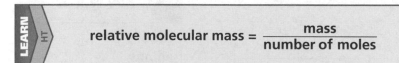

LEARN HT
$$\text{relative molecular mass} = \frac{\text{mass}}{\text{number of moles}}$$

What is the relative molecular mass of magnesium hydroxide, $Mg(OH)_2$?

Mg: 1×24	=	24
O: 2×16	=	32
H: 2×1	=	2
M_r: $24 + 32 + 2$	=	58

The formula has been given: $Mg(OH)_2$

> **HT Key Point**
>
> Carbon-12 is the pure isotope of carbon, which has the atomic mass of precisely 12.

> **HT Key Point**
>
> 6.022×10^{23} is written in standard form notation because writing 602 200 000 000 000 000 000 000 is extremely awkward.

The relative formula mass of $Mg(OH)_2$ is 58, so the relative molecular mass of $Mg(OH)_2$ is 58g/mol.

How many moles of ethanol are there in 230g of ethanol? (The relative formula mass of ethanol is 46.)

$$\text{number of moles} = \frac{\text{mass}}{\text{relative molecular mass}}$$

$$= \frac{230g}{46g/mol} = 5mol$$

- If the mass of one mole of a chemical is known, then the mass of one atom or molecule can be worked out.

One mole of sulfur has a mass of 32g.
What is the mass of one sulfur atom?

$$\frac{\text{atomic mass of element}}{\text{Avogadro's constant}} = \frac{32g}{6.022 \times 10^{23}} = 5.3 \times 10^{-23}g$$

 Key Point

Showing the units in an equation helps because they cancel out. If the final unit matches what you are trying to find out, you have done the calculation correctly.

Calculating Masses of Reactants or Products

- The ratio of the experimental mass to the atomic mass of the constituent atoms can be used to predict the amount of product in a reaction or vice versa.

How much water will be produced when 2 moles of hydrogen is completely combusted in air?

$2H_2(g) + O_2(g) \rightarrow 2H_2O(l)$

relative molar mass of water = $(2 \times 1) + 16 = 18g/mol$
mass of water produced = $2 \times 18 = 36g$

2 moles of hydrogen produce 2 moles of water.

72g of water is produced in the same reaction, how much oxygen was reacted?

$2H_2(g) + O_2(g) \rightarrow 2H_2O(l)$

relative molecular mass of water = $(2 \times 1) + 16 = 18g/mol$
relative molecular mass of oxygen = $2 \times 16 = 32g/mol$

moles of water produced = $\frac{72}{18} = 4mol$

moles of oxygen used = $2mol$
mass of oxygen used = $2 \times 32 = 64g$

Since 2 moles of water are formed from 1 mole of oxygen, divide by 2.

 Quick Test

1. Write the equation that you would use to work out mass from the relative molecular mass and number of moles.
2. 16g of oxygen reacts fully with hydrogen. How much water is produced?
3. The relative atomic mass of caesium (Cs) is 133. What is the mass of a single atom?

Key Words

mole
carbon-12
Avogadro's constant
relative atomic mass
relative molecular mass

Energetics

You must be able to:

- Explain the difference between endothermic and exothermic reactions
- Draw and label reaction profiles for an endothermic and an exothermic reaction
- **HT** Calculate energy changes in a chemical reaction considering bond energies.

Reactions and Temperature

- In a chemical reaction, energy is taken in or given out to the surroundings.
- **Exothermic** reactions release energy to the surroundings causing a temperature rise, e.g. when wood burns through combustion.
- The energy given out by exothermic chemical reactions can be used for heating or to produce electricity, sound or light.
- **Endothermic** reactions absorb energy from the surroundings and cause a temperature drop.
- For example, when ethanoic acid (vinegar) and calcium carbonate react, the temperature of the surroundings decreases.
- Endothermic reactions can be used to make cold packs, which are used for sports injuries.

Activation Energy

- Most of the time chemicals do not spontaneously react.
- A minimum amount of energy is needed to start the reaction. This is called the **activation energy**.
- For example, paper does not normally burn at room temperature.
- To start the combustion reaction, energy has to be added in the form of heat from a match. This provides enough energy to start the reaction.
- As the reaction is exothermic, it will produce enough energy to continue the reaction until all the paper has reacted (burned).

Reaction Profiles

- A graph called a **reaction profile** can be drawn to show the energy changes that take place in exothermic and endothermic reactions.

HT Energy Change Calculations

- In a chemical reaction:
  - making bonds is an exothermic process (releases energy)
  - breaking bonds is an endothermic process (requires energy).
- Chemical reactions that release more energy by making bonds than breaking them are exothermic reactions.

> **Key Point**
>
> Energy is never lost or used up, it is just transferred.

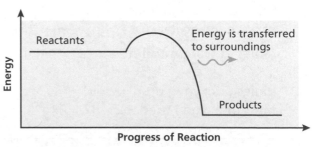

Reaction Profile for an Exothermic Reaction

Reactants

Energy is transferred to surroundings

Products

Energy

Progress of Reaction

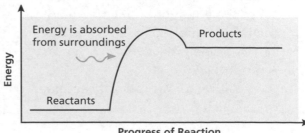

Reaction Profile for an Endothermic Reaction

Energy is absorbed from surroundings

Products

Reactants

Energy

Progress of Reaction

- That energy was originally stored in the bonds between atoms in the reactants.
- Chemical reactions that need more energy to break bonds than is released when new bonds are made are endothermic reactions.
- The energy taken in from the **environment** is converted to **bond energy** between the atoms in the products.
- To work out whether a reaction is exothermic or endothermic, calculations can be carried out using information about how much energy is released when a bond forms and how much energy is needed to break a bond.
- The steps to follow are:
  1 Write out the balanced equation and look at the bonds.
  2 Add up the energies associated with breaking bonds in the reactant(s).
  3 Add up the energies associated with making bonds in the product(s).
  4 Calculate the energy change using the equation below:

> energy change = energy used to break bonds – energy released when new bonds are made

- If the energy change is negative, the reaction is exothermic (more energy is released making bonds than is used breaking them).
- If the energy change is positive, the reaction is endothermic (less energy is released making bonds than is used breaking them).

Hydrogen reacts with iodine to form hydrogen iodide. Calculate the energy change for this reaction.

Bond	Bond Energy (kJ/mol)
H–H	436
I–I	151
H–I	297

$H_2(g) + I_2(g) \rightarrow 2HI(g)$

Total energy needed to break the bonds in
the reactants   = 436 + 151
                  = 587kJ/mol

Total energy released making the bonds in
the product   = 2 × 297
                  = 594kJ/mol

Energy change   = 587 – 594
                  = –7kJ/mol

Key Point

In the exam, you will be given the bond energy values. You do not have to memorise them.

The reactants contain one H–H bond and one I–I bond. The products contain two H–I bonds.

Energy change is negative, so the reaction is exothermic.

Quick Test

1. What is 'activation energy'?
2. Draw a reaction profile for an endothermic reaction.
3. **HT** The bond making and bond breaking energies in a chemical reaction add up to –15kJ/mol. Is the reaction exothermic or endothermic?

Key Words

exothermic
endothermic
activation energy
reaction profile
 environment
 bond energy

Types of Chemical Reactions

You must be able to:

- Explain whether a substance is oxidised or reduced in a reaction
- **HT** Explain oxidation and reduction in terms of loss and gain of electrons
- Predict the products of reactions between metals or metal compounds and acids.

Oxidation and Reduction

- When oxygen is added to a substance, it is **oxidised**.
- When oxygen is removed from a substance, it is **reduced**.
- The substance that gives away the oxygen is called the **oxidising agent**.
- The substance that receives the oxygen is the **reducing agent**.

> **copper oxide + hydrogen ⟶ copper + water** ◄

Copper oxide is the oxidising agent (it loses the oxygen). Hydrogen is the reducing agent (it gains the oxygen to form water).

HT Loss and Gain of Electrons

- Chemists modified the definition of oxidation and reduction when they realised that substances could be oxidised and reduced without oxygen being present.
- The definition now focuses on the loss or gain of electrons in a reaction:
  - If a substance gains electrons, it is reduced.
  - If a substance loses electrons, it is oxidised.

> **$2Na(s) + Cl_2(g) \longrightarrow 2NaCl(s)$** ◄

HT Key Point

OILRIG: Oxidation **I**s **L**oss (of electrons), **R**eduction **I**s **G**ain (of electrons).

Sodium gives away the single electron in its outermost shell, so it has been oxidised. Chlorine receives the electrons from the two sodium atoms, so it has been reduced.

Acids and Alkalis

- When an acid or alkali is dissolved in water, the ions that make up the substance move freely.
  - An **acid** produces hydrogen ions, $H^+(aq)$.
  - An **alkali** produces hydroxide / hydroxyl ions, $OH^-(aq)$.
- For example, a solution of hydrochloric acid, HCl, will dissociate into $H^+(aq)$ and $Cl^-(aq)$ ions.
- A solution of sodium hydroxide, NaOH, will dissociate into $Na^+(aq)$ and $OH^-(aq)$ ions.

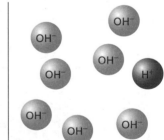

low pH = lots of H^+ lots of OH^- = high pH

Neutralisation

- **Neutralisation** occurs when an acid reacts with an alkali or a **base**, to form a **salt** and water.

> **acid + base ⟶ salt + water**

- For example, hydrochloric acid reacts with sodium hydroxide to produce sodium chloride and water:

$$\text{HCl(aq) + NaOH(aq)} \longrightarrow \text{NaCl(aq) + H}_2\text{O(l)}$$

- The reaction can be rewritten to only show the species that change:

$$\text{H}^+\text{(aq) + OH}^-\text{(aq)} \longrightarrow \text{H}_2\text{O(l)}$$

Reacting Metals with Acid

- Many metals will react in the presence of an acid to form a salt and hydrogen gas.

LEARN

> **metal + acid** $\longrightarrow$ **salt + hydrogen**

- The reactivity of a metal determines whether it will react with an acid and how vigorously it reacts.
- Metals can be arranged in order of reactivity in a reactivity series.
- If there is a reaction, then the name of the salt produced is based on the acid used:
  - Hydrochloric acid forms chlorides.
  - Nitric acid forms nitrates.
  - Sulfuric acid forms sulfates.

magnesium + hydrochloric acid $\longrightarrow$

$$\text{magnesium chloride + hydrogen}$$
$$\text{Mg(s) + 2HCl(aq)} \longrightarrow \text{MgCl}_2\text{(aq) + H}_2\text{(g)}$$

Reacting Metal Carbonates with Acid

- Metal carbonates also react with acids to form a metal salt, plus water and carbon dioxide gas.

LEARN

> **metal carbonate + acid** $\longrightarrow$
> **salt + water + carbon dioxide**

- The salts produced are named in the same way as for metals reacting with acids.

magnesium carbonate + sulfuric acid $\longrightarrow$

$$\text{magnesium sulfate + water + carbon dioxide}$$
$$\text{MgCO}_3\text{(s) + H}_2\text{SO}_4\text{(aq)} \longrightarrow \text{MgSO}_4\text{(aq) + H}_2\text{O(l) + CO}_2\text{(aq)}$$

Key Point

Remember, ionic substances separate from each other when dissolved or molten. The ions move freely and are not joined together.

Key Point

Water is not an ionic compound. It is a polar molecule (it has positively charged hydrogen and negatively charged oxygen), which means that ionic substances can dissolve easily into it.

Reactivity Series

The higher the metal is positioned the more readily it reacts with oxygen. This is useful for protecting metals lower down against corrosion. →

These metals slowly react with oxygen and corrode away. →

This metal will very slightly discolour to show oxygen has had very little effect. It very rarely corrodes. →

These metals remain unaffected by oxygen. →

Most Reactive
Sodium
Calcium
Magnesium
Aluminium
Zinc
Iron
Lead
Copper
Gold
Platinum
Least Reactive

Key Words

oxidised
reduced
oxidising agent
reducing agent
acid
alkali
neutralisation
base
salt

Quick Test

1. What gas is made when metal carbonates react with acid?
2. What salt is made when zinc oxide is reacted with nitric acid?
3. Write the word equation for the reaction between copper oxide and sulfuric acid.

pH, Acids and Neutralisation

You must be able to:

* Describe techniques to measure pH
* HT Explain the terms dilute, concentrated, weak and strong in relation to acids
* HT Explain pH in terms of dissociation of ions.

Measuring pH

* Indicators change colour depending on whether they are in acidic or alkaline solutions.
* Single indicators, such as litmus, produce a sudden colour change when there is a change from acid to alkali or vice versa.
* **pH** is a scale from 0 to 14 that provides a measure of how acidic or alkaline a solution is.
* Universal indicator is a mixture of different indicators, which gives a continuous range of colours.
* The pH of a solution can be estimated by comparing the colour of the indicator in solution to a pH colour chart.

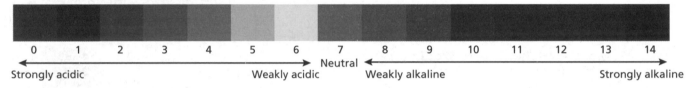

| 0   1   2   3   4   5   6   7   8   9   10   11   12   13   14 |

Neutral

Strongly acidic   Weakly acidic   Weakly alkaline   Strongly alkaline

* pH can also be measured electronically using an electronic data logger with a pH probe, which gives the numerical value of the pH.

 Key Point

Judging something using the eye is a qualitative measurement and has more variation than a quantitative measurement, such as a pH reading from a pH probe.

HT Dilute and Concentrated Acids

* Acids can be **dilute** or **concentrated**.
* The degree of dilution depends upon the amount of acid dissolved in a volume of water.
* The higher the ratio of acid to water in a solution, the higher the concentration.
* Acids **dissociate** (split apart) into their component ions when dissolved in solution.
* The concentration is measured as the number of moles of acid per cubic decimetre of water (mol/dm³).
* For example, 1mol/dm³ is less concentrated than 2mol/dm³ of the same acid.

Key Point

Don't confuse the term 'concentrated' with how 'strong' an acid or alkali is.

HT Strong and Weak Acids

* The terms **weak acid** and **strong acid** refer to how well an acid dissociates into ions in solution.
* Strong acids easily form H^+ ions.

$$HCl(aq) \longrightarrow H^+(aq) + Cl^-(aq)$$
$$HNO_3(aq) \longrightarrow H^+(aq) + NO_3^-(aq)$$
$$H_2SO_4(aq) \longrightarrow 2H^+(aq) + SO_4^{2-}(aq)$$

- These strong acids fully ionise.
- Acids that do not fully ionise form an **equilibrium mixture**.
- This means that the ions that are formed can recombine into the original acid. For example:

$$\text{ethanoic acid} \rightleftharpoons \text{ethanoate ions} + \text{hydrogen ions}$$
$$CH_3COOH(aq) \rightleftharpoons CH_3COO^-(aq) + H^+(aq)$$

HT Changing pH

- pH is a measure of how many hydrogen ions are in solution.
- Changing the concentration of an acid leads to a change in pH.
- The more concentrated the acid, the lower the pH and vice versa.
- The concentration of hydrogen ions will be greater in a strong acid compared to a weak acid.
- The pH of a strong acid will therefore be lower than the equivalent concentration of weak acid.
- As the concentration of H^+ ions increases by a factor of 10, the pH decreases by one unit.
- A solution of an acid with a pH of 4 has 10 times more H^+ ions than a solution with a pH of 5.
- A solution of an acid with a pH of 3 has 100 times more H^+ ions than a solution with a pH of 5.

Concentrated weak acid – a lot of acid present, but little dissociation of acid

Concentrated strong acid – a lot of acid present with a lot of dissociation to form many hydrogen ions

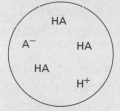

$$\text{acid (HA)} \rightleftharpoons \text{hydrogen ion } (H^+) + \text{anion } (A^-)$$

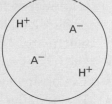

Dilute weak acid – little acid present with little dissociation of acid

Dilute strong acid – little acid present but with a high degree of dissociation

HT Neutralisation and pH

- For neutralisation to occur, the number of H^+ ions must exactly cancel the number of OH ions.
- pH curves can be drawn to show what happens to the pH in a neutralisation reaction:
  - An acid has a low pH – when an alkali is added to it, the pH increases.
  - An alkali has a high pH – when an acid is added to it, the pH decreases.
- You should be able to read and interpret pH curves (like the one opposite) to work out:
  - the volume of acid needed to neutralise the alkali
  - the pH after a certain amount of acid has been added.

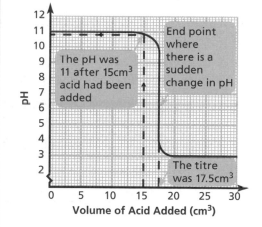

The pH was 11 after 15cm³ acid had been added

End point where there is a sudden change in pH

The titre was 17.5cm³

Quick Test

1. Why can universal indicator be more useful than litmus indicator?
2. HT What two pieces of information can a pH curve tell you about an acid or alkali?
3. HT What is pH a measure of?

Electrolysis

You must be able to:

- Predict the products of electrolysis of simple ionic compounds in the molten state
- Describe the competing reactions in the electrolysis of aqueous compounds
- Describe electrolysis in terms of the ions present and the reactions at the electrodes
- Describe the technique of electrolysis using inert and non-inert electrodes

Electrolysis

- Ionic compounds can be broken down into their constituent elements using electricity. The substance being broken down is known as the **electrolyte**. The electrolyte must be molten or dissolved in water so that the ions can move and conduct electricity.
- Electrodes are made of solid materials that conduct electricity.
- The positively charged electrode is called the **anode**.
- The negatively charged electrode is called the **cathode**.
- During **electrolysis**, **cations** (positively charged ions) are attracted to the cathode and **anions** (negatively charged ions) to the anode.

Electrolysis of Molten Compounds, e.g. NaCl

- During the electrolysis of molten sodium chloride, the cations (sodium ions) are attracted to the cathode. Here they gain electrons and turn into sodium atoms.
- Metallic sodium can be seen to form at the cathode.

HT This is a reduction reaction. A reduction reaction occurs when a species gains electrons.

HT This process can be shown by writing a half-equation.

> **HT At the cathode:**
> $$Na^+ + e^- \longrightarrow Na$$

- The anions (chloride ions) are attracted to the anode. Here each chloride ion loses an electron, to form a chlorine atom. Two chlorine atoms pair up to form a chlorine molecule.

HT This is an oxidation reaction. An oxidation reaction occurs when a species loses electrons.

HT The half-equation for this reaction is

> **HT At the anode:**
> $$2Cl^- \longrightarrow Cl_2 + 2e^-$$

- Chlorine gas can be seen to form at the anode.
- The electrons produced at the anode are pumped by the battery through the wires in the circuit to the cathode, where they are given to the sodium ions.

> ### Key Point
>
> Unless the ions can move (i.e. the substance is in solution or molten) electrolysis will not occur.

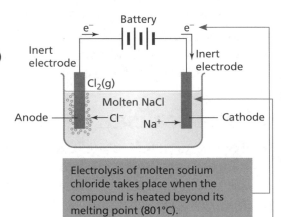

Electrolysis of molten sodium chloride takes place when the compound is heated beyond its melting point (801°C).

Na(s) forming on surface of cathode.

This is a reduction process as electrons are gained.

This is an oxidation process as electrons are lost.

- The products of molten binary ionic compounds (i.e. ionic compounds made up of two elements) will always be the two elements present in the compound. The metal will always be formed at the cathode and the non-metal at the anode.

Electrolysis of Aqueous Solutions, e.g. CuSO₄ (aq)

- Aqueous solutions contain cations and anions from the ionic compound dissolved in the water.
- They also contain H^+ ions and OH^- ions from the water.
- This means the ions shown in the table alongside are present in copper sulfate solution.
- Only one ion is attracted to each electrode.
- At the cathode 'the least reactive element is formed'.
- Copper is below hydrogen in the reactivity series and so will be formed at the cathode.

> **HT** The half-equation for this reaction is $Cu^{2+} + 2e^- \longrightarrow Cu$

- When inert (non-reactive) electrodes are used, the product at the cathode is always a metal or hydrogen (if hydrogen is less reactive than the metal that is also present).
- At the anode 'oxygen is formed unless a halogen (group 7) ion is present'.
- In the electrolysis of copper sulfate solution, there are no halogen ions present so oxygen is formed at the anode.

> **HT** The half-equation for this reaction is $4OH^- \longrightarrow O_2 + 2H_2O + 4e^-$

- The H^+ ions and SO_4^{2-} ions are unaffected and remain in solution.
- Non-metals are always formed at the anode when inert electrodes are used.

Use of Inert and Non-Inert Electrodes

- **Inert electrodes** do not react during electrolysis.
- Typically they are made from carbon.
- Electrodes can be made out of inert metals instead, such as platinum, which will not react with the products of electrolysis. But, platinum electrodes are very expensive.
- Non-inert or **active electrodes** can be used for processes such as electroplating, e.g. using copper electrodes with copper sulfate solution.
- In the electrolysis here, if the cathode were replaced with a metal object it would become covered in copper metal, i.e. it will be copper-plated.

> **Quick Test**
>
> 1. Name the products of electrolysis of **molten** magnesium bromide.
> 2. Name the products of electrolysis of **aqueous** magnesium bromide.
> 3. **HT** During the electrolysis of molten aluminium oxide, aluminium is formed from aluminium ions. Write a half-equation to show this reaction.

Electrolysis of Copper Sulfate Solution

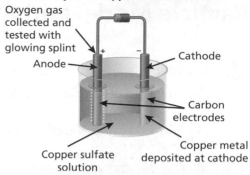

Oxygen gas collected and tested with glowing splint

Anode

Cathode

Carbon electrodes

Copper metal deposited at cathode

Copper sulfate solution

Cations Present	Anions Present
Cu^{2+}	SO_4^{2-}
H^+	OH^-

> ### Key Point
>
> Ionic solutions conduct electricity because the ions that make up the solution move to the electrodes, *not* because electrons move through the solution.

Active Electrodes

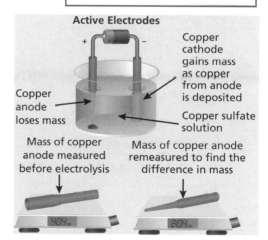

Copper cathode gains mass as copper from anode is deposited

Copper anode loses mass

Copper sulfate solution

Mass of copper anode measured before electrolysis

Mass of copper anode remeasured to find the difference in mass

> ### Key Words
>
> electrolyte
> anode
> cathode
> electrolysis
> cations
> anions
> inert electrode
> active electrode

Review Questions

Particle Model and Atomic Structure

1 Tad places a cup of water on a windowsill.
When he returns four days later, he notices that all of the water has disappeared.

   a) What has happened to the water? [1]

   b) Tad decides place another cup on the windowsill.

      What would Tad need to do to increase the rate of water loss? [1]

   c) Draw two boxes:

      i) In one box show the arrangement of the water particles in the cup. [1]

      ii) In the other box, draw the arrangement of water particles in the air. [1]

2 Which of the following is the approximate **mass** of an atom?

   A 1×10^{-1}g   **B** 1×10^{-10}g   **C** 1×10^{-15}g   **D** 1×10^{-23}g   [1]

3 Complete the table showing the structure of an atom. [5]

Subatomic Particle	Relative Mass	Relative Charge
	1	
Neutron		0 (neutral)
Electron		

4 In 1911, Geiger and Marsden fired alpha particles at gold foil.
They expected all of the alpha particles to pass straight through the atoms.
Instead, a substantial number bounced straight back.

   a) What subatomic particle must have been present in the middle of the atom for this to happen? [1]

   b) Suggest why only some of the alpha particles bounced back. [2]

5 For each of the following elements, work out how many **neutrons** are present:

2
He
helium
4.0

11
Na
sodium
23.0

23
V
vanadium
50.9

[3]

Total Marks _____ / 16

Purity and Separating Mixtures

1 Sanjit is testing a mystery pure substance.
He heats the substance until it boils.

   a) How could he confirm the identity of the substance?   [2]

   b) Sanjit now takes a sample of water and boils it.
      He finds that the water boils at 101°C.

      Assuming that the thermometer is working correctly, why does the water boil at
      this temperature?   [2]

2 What is the empirical formula of each of the following substances?

   a) $C_6H_{12}O_6$ [1]

   b) CH_3COOH [1]

   c) $CH_3CH_2CH_2COOH$ [1]

3 Calculate the relative formula masses of the following compounds:

   a) $C_6H_{12}O_6$ [1]   **b)** CH_3COOH [1]

   c) CO_2 [1]   **d)** H_2SO_4 [1]

4 What is the scientific term for a substance created by mixing two or more elements, at least
one of which is a metal?   [1]

5 Three substances are separated using thin layer chromatography.

Substance	R_f Value
A	0.6
B	0.3
C	0.7

Which substance travelled the furthest in the solid stationary phase?   [1]

Total Marks _____ / 13

Review Questions

Bonding

1 Which of the following scientists first devised the structure of the modern periodic table? Circle the correct answer.

Bohr   Mendel   Mendeleev   Marsden   Rutherford   Thomson   [1]

2 Here are the electronic structures of five elements:

Element	Electronic Structure
A	2.1
B	2.8.1
C	2.8.3
D	2.8.8
E	2.2

a) Which elements are in Period 2 of the periodic table? [2]

b) Which elements are in the same group? [2]

c) Which is the most reactive metal? [1]

d) What is the atomic number of each element? [5]

3 The majority of elements in the periodic table are metals.

a) Give **three** properties of metals. [3]

b) What is the general name given to the product formed when a metal reacts with oxygen? [1]

c) What type of compound is formed when metals react with non-metals? [1]

Total Marks _____ / 16

Models of Bonding

1 Which of the following diagrams correctly shows the bonding for fluorine gas (F_2):

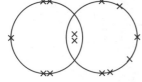

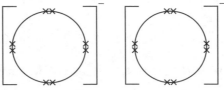

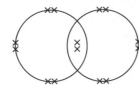

2. Draw a dot and cross diagram to show the covalent bonds in **methane** (CH_4). [2]

3. What advantage do ball and stick models have over dot and cross diagrams? [2]

4. Show, using dot and cross diagrams, how magnesium becomes a magnesium ion. [2]

[1]

Total Marks _____ / 7

Properties of Materials

1. Carbon has a number of different **allotropes**.

   a) What is meant by the term **allotrope**? [1]

   b) Graphite is used as a dry lubricant in air compressors.

   Referring to the structure of graphite, explain why it is a good dry lubricant. [4]

   c) Diamonds are often used in drill bits.

   Referring to the structure of diamond, explain why diamonds are used in drill bits. [3]

   d) Give **one** use of graphene. [1]

2. Silver nanoparticles are often impregnated into items such as plastic chopping boards and sticking plasters.

   a) Suggest why silver nanoparticles are impregnated into these items. [1]

   b) Why are **nanoparticles** so named? [1]

   c) Explain why some people are concerned about the risks of using nanoparticles. [2]

3. Which property of nanoparticles makes them suitable for use as a catalyst?
   Circle the correct answer: [1]

   sized between 1 and 100mm   **small surface area**   **large surface area**   **unreactive**

Total Marks _____ / 14

Practice Questions

Introducing Chemical Reactions

1 What is the **law of conservation of mass**? [1]

2 **a)** What do the **subscript** numbers that appear after an element symbol mean, e.g. Cl_2? [1]

b) Write the number of atoms of each element shown in each formula below:

    i) $C_6H_{12}O_6$ [1]

    ii) CH_3CH_2COOH [1]

    iii) H_2O_2 [1]

    iv) $Ca(NO_3)_2$ [1]

3 Write down the four state symbols. [1]

4 Write the **balanced symbol equations** for the following reactions, including state symbols:

a) magnesium + oxygen → magnesium oxide [2]

b) lithium + oxygen → lithium oxide [2]

c) calcium carbonate + hydrochloric acid → calcium chloride + carbon dioxide + water [2]

d) aluminium + oxygen → aluminium oxide [2]

Total Marks .... / 15

Chemical Equations

1 What are the charges on these common ions?

a) copper(II) [1]

b) oxide [1]

c) iron(III) [1]

d) sulfide [1]

2 HT Write the half equation for each of the following reactions:

a) Hydrogen ions to hydrogen gas $2H^+ + 2e^- \rightarrow H_2$ [1]

b) Iron(II) ions to iron solid $Fe^{2+} + 2e^- \rightarrow Fe$ [1]

c) Copper(II) ions to copper solid $Cu^{2+} + 2e^- \rightarrow Cu$ [1]

d) Zinc to zinc ions $Zn \rightarrow Zn^{2+} + 2e^-$ [1]

3 HT Write the ionic equation for the following reaction.
All the compounds involved are soluble, except for silver chloride. [2]

silver nitrate + lithium chloride → lithium nitrate + silver chloride

Total Marks _____ 8 _____ / 10

Moles and Mass

1 HT What does a **mole** represent in chemistry? [1]

2 HT Which of the following is Avogadro's constant?

A 6.022×10^{32} C 3.142×10^{32}

B 6.022×10^{23} D 3.142×10^{23} [1]

3 HT What unit is **molecular mass** measured in? g/mol [1]

4 HT Cyanobacteria are organisms that can convert atmospheric nitrogen into nitrates.
Abigail is preparing stock solutions containing different metals to investigate how they affect the growth of cyanobacteria.

42	23
Mo	**V**
molybdenum	vanadium
95.9	50.9

She weighs out 287.7g of the element molybdenum.

a) How many moles of molybdenum does she have? $\frac{287.7}{95.9} = 3$
Show your working. [2]

b) Abigail needs to weigh out 5 moles of vanadium.

What mass of vanadium should she use? $50.9 \times 5 = 254.5$
Show your working. [2]

5 HT What is the relative molecular mass of glucose, $C_6H_{12}O_6$?
(Relative atomic mass of C = 12, H = 1 and O = 16.) [1]

6 HT Calculate the mass of one atom of each of the following elements.
Show your working. Give your answer to one decimal place.

a) Vanadium [2]

b) Molybdenum [2]

c) Caesium [2]

d) Bismuth [2]

7 HT Five moles of hydrogen are completely combusted in air.

How much water is produced in the reaction? [2]

Total Marks _____ / 18

Energetics

1 a) A reaction gives out energy to the environment.

What type of reaction is it? [1]

b) A reaction takes in energy from the environment.

What type of reaction is it? [1]

2 Give **two** ways in which the energy released from a reaction can be used. [2]

3 Explain what is meant by the term **activation energy**. [1]

4 Draw a reaction profile for an exothermic reaction. [1]

5 Draw a reaction profile for an endothermic reaction. [1]

6 HT Which of the following is an **exothermic** process?

making chemical bonds   breaking chemical bonds [1]

7 HT Peter reacted hydrogen gas with fluorine gas to form hydrogen fluoride gas.

The equation for the reaction is: $H_2(g) + F_2(g) \rightarrow 2HF(g)$

Bond	Bond Energy (kJ/mol)
H–F	565
H–H	432
F–F	155

Calculate the energy change for this reaction and state whether the reaction is **exothermic** or **endothermic**. [4]

8 HT A series of reactions was carried out and the energy changes were recorded.

For each energy change, state whether it was **exothermic** or **endothermic**.

a) +90kJ/mol [1]

b) –181kJ/mol [1]

c) +20kJ/mol [1]

d) +8kJ/mol [1]

9 HT Look at the reaction profile for $H_2(g) + F_2(g) \rightarrow 2HF(g)$

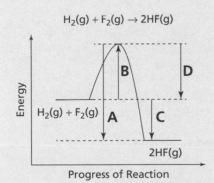

Which letter on the diagram shows the energy change for the reaction? [1]

Total Marks _____ / 17

Practice Questions

Types of Chemical Reactions

1 Which reactions involve a reactant being oxidised? [2]

    A magnesium + oxygen → magnesium oxide

    B water (solid) → water (liquid)

    C copper + oxygen → copper oxide

    D barium carbonate + sodium sulfate → barium sulfate + sodium carbonate

2 HT Explain what oxidation and reduction mean in terms of electrons. [2]

3 For each of the following reactions, write a balanced equation, including state symbols. Then state which species has been oxidised and which has been reduced.

    a) sodium + chlorine → sodium chloride [3]

    b) magnesium + oxygen → magnesium oxide [3]

    c) lithium + bromine → lithium bromide [3]

    d) copper(II) oxide + hydrogen → copper + water [3]

4 What ions are produced by:

    a) An acid? [1]

    b) An alkali? [1]

5 What is the general equation for the neutralisation of a base by an acid? [1]

6 Dilute sulfuric acid and sodium hydroxide solution are reacted together.

    a) Write the balanced symbol equation for the reaction. [2]

    b) Which ions are not involved in the reaction? [2]

    c) HT Write the ionic equation for the reaction between dilute sulfuric acid and sodium hydroxide solution. [2]

Total Marks _____ / 25

pH, Acids and Neutralisation

1 HT What is meant by the term **weak acid**? [1]

2 HT Look at the concentrations below. For each pair, which is more concentrated?

    a) 1mol/dm³ H_2SO_4 OR   2mol/dm³ H_2SO_4 [1]

    b) 3mol/dm³ HNO_3 OR   2mol/dm³ HNO_3 [1]

3 HT How many more times concentrated are the H^+ ions in a solution with a pH of 6 compared to a solution with a pH of 3? [1]

Total Marks .... / 4

Electrolysis

1 What are the ions of **a)** metals and **b)** non-metals called? [2]

2 What is **electrolysis**? [1]

3 Why is it not possible to carry out electrolysis on crystals of table salt (sodium chloride) at room temperature and pressure? [1]

4 Describe how you could copper-plate a nail using copper(II) sulfate solution. [3]

5 **a)** Why are inert electrodes often used in electrolysis? [1]

    b) Platinum can be used as an inert electrode. However, they are rarely used.

    Why are platinum electrodes rarely used? [1]

Total Marks .... / 9

Predicting Chemical Reactions

You must be able to:

- Describe the properties of elements in Group 1, Group 7 and Group 0
- Recall the general properties of transition metals
- Predict possible reactions of elements from their position in the periodic table
- Work out the order of reactivity of different metals.

Group 1

- Group 1 metals all have one **electron** in their outer shell.
- The outer shell can take a maximum of eight electrons (for elements 3–20) or two for H and He.
- Elements are **reactive** because atoms will gain or lose electrons until they have a full outer shell.
- The first three elements in the group are lithium (Li), sodium (Na) and potassium (K).
- It is easier for a Group 1 metal to donate its outermost electron than gain seven more electrons to achieve a full outer shell.
- The elements in Group 1 have similar physical and chemical properties.
- Density increases going down the group.
- The first three metals float on water.
- The Group 1 metals become more reactive going down the group.
- This is because the outermost electron gets further away from the nucleus, so the force of attraction between the electron and nucleus is weaker.

Group 7

- The non-metals in Group 7 are known as the **halogens**.
- They all have seven electrons in their outer shell, so they have similar chemical properties.
- It is easier for a Group 7 halogen to gain a single electron than donate seven electrons to achieve a full outer shell.
- Fluorine (F), chlorine (Cl), bromine (Br) and iodine (I) are halogens.
- The elements in Group 7 have similar physical and chemical properties.
- Their melting and boiling points are very low and they are not very dense.
- Unlike the alkali metals, halogens become less reactive going down the group.
- This is because the closer the outer shell is to the nucleus, the stronger the force of attraction, and the easier it is to gain an electron.

Group 0

- The non-metals in Group 0 are known as the **noble gases**.
- These elements all have a full outer shell and do not react with other elements.

Key Point

It is wrong to say that a Group 1 metal 'wants' to lose an electron. There is a force of attraction between the electrons (negative) and the nucleus (positive). The atom cannot hold onto its outermost electron if there is a stronger force of attraction present.

Group 1

Li	lithium
Na	sodium
K	potassium

Reactivity increases, and melting and boiling points decrease as you go down the group.

Group 7

F	fluorine
Cl	chlorine
Br	bromine
I	iodine

Reactivity decreases, and melting and boiling points increase as you go down the group.

Transition Metals

- The **transition metals** are a block of metallic elements between Groups 2 and 3 of the periodic table.
- Transition metals are used as catalysts in chemical reactions.
- They form coloured ions with different charges.
- They tend to have high melting points, are dense, and are not as reactive as the metals in Groups 1 and 2.
- The transition metals have different numbers of electrons in their outer shells.

Predicting Reactivity

- In general, for metals:
  - the lower down the group, the more reactive the element
  - the fewer the electrons in the outer shell, the more reactive it will be.
- For example, potassium (K) and calcium (Ca) are in the same period on the periodic table.
- K is more reactive than Ca because Ca has two electrons in its outer shell and K only has one.
- All metals can be arranged in a **reactivity series**.
- A metal higher in the reactivity series can displace another metal from a compound, e.g. copper is lower than magnesium in the reactivity series, so it will be displaced by magnesium.

$$CuSO_4(aq) + Mg(s) \longrightarrow MgSO_4(aq) + Cu(s)$$

- In general, for non-metals:
  - the higher up the group, the more reactive the element
  - the greater the number of electrons in the outer shell, the more reactive it will be.

Metal Reactions with Acids and Water

- Very reactive metals will react with acid to form metal salts + hydrogen:

$$metal + acid \longrightarrow metal\ salt + hydrogen$$

- Reactive metals will react with water to form metal hydroxides + hydrogen.
- The more reactive the metal, the quicker it will donate its outer electron(s).
- Positive metal ions are formed.

$$magnesium + water \longrightarrow magnesium\ hydroxide + hydrogen$$
$$Mg(s) + 2H_2O(l) \longrightarrow Mg(OH)_2(aq) + H_2(g)$$

Quick Test

1. Which of these halogens would be the least reactive: bromine, chlorine, fluorine or iodine?
2. Explain why caesium is more reactive than lithium.
3. Which of the following metals will displace copper from $CuSO_4$: silver, platinum, lithium?

Reactivity Series

Most Reactive
Sodium
Calcium
Magnesium
Aluminium
Carbon
Zinc
Iron
Lead
Hydrogen
Copper
Gold
Platinum
Least Reactive

Key Point

The reactivity series also includes hydrogen and carbon. Although non-metals, these elements also displace metals.

$Mg(OH)_2$ is an ionic compound so it will dissociate into $Mg^{2+}(aq)$ and $OH^-(aq)$ ions.

Key Words

electron
reactive
halogen
noble gas
transition metal
reactivity series

Identifying the Products of Chemical Reactions

You must be able to:

- Describe the tests to identify gases produced in a chemical reaction
- Describe the tests to identify metal cations in solution
- Describe the tests to identify non-metal anions in solution.

Identifying Gases

- If bubbles or fizzing are observed during a reaction, then a gas has been produced.
- If you have a chemical equation for a reaction you should know the name of the gas.
- However, to prove the identity of a gas you must carry out an experimental test.

Key Point

It is important to be very clear about what you are seeing when carrying out gas tests. Making comparisons before and after is useful.

Gas	Properties	Test for Gas
Oxygen, O_2	A colourless gas that helps fuels burn more readily in air.	Relights a glowing splint.
Carbon dioxide, CO_2	A colourless gas produced when fuels are burned in a sufficient supply of oxygen.	Turns limewater milky.
Hydrogen, H_2	A colourless gas. It combines violently with oxygen when ignited.	When mixed with air, burns with a squeaky pop.
Chlorine, Cl_2	A green poisonous gas that bleaches dyes.	Turns damp indicator paper white.

Testing for Aqueous Ions

- A **displacement** reaction occurs when a metal that is higher in the **reactivity series** displaces another metal from its compound.
- In metal displacement reactions, a **precipitate** (a solid) will be formed.

Metal Cations

- All transition metals, as well as calcium from Group 2, will form a precipitate when reacted with sodium hydroxide solution.
- The metal ions form an insoluble precipitate of a metal hydroxide.
- The precipitate formed often has a characteristic colour.
- This means that it is possible to identify which metal ions were in solution.
- For example, calcium, copper, iron(II), iron(III) and zinc cations all react with the hydroxide anions.

Key Point

Scientists have to use a range of different experimental methods to determine what chemicals are present. It is important to keep clear records of the results and what techniques were used to obtain them.

Calcium ions: calcium ions + hydroxide ions ⟶ calcium hydroxide (a white precipitate)

HT $$Ca^{2+}(aq) + 2OH^-(aq) \longrightarrow Ca(OH)_2(s)$$

Copper ions: copper + hydroxide ions ⟶ calcium hydroxide (a light blue precipitate)

HT $$Cu^{2+}(aq) + 2OH^-(aq) \longrightarrow Ca(OH)_2(s)$$

Iron(II) ions: iron + hydroxide ions ⟶ iron(II) hydroxide (a green precipitate)

HT $$Fe^{2+}(aq) + 2OH^-(aq) \longrightarrow Fe(OH)_2(s)$$

Iron(III) ions: iron + hydroxide ions ⟶ iron(III) hydroxide (a red-brown precipitate)

HT $$Fe^{3+}(aq) + 3OH^-(aq) \longrightarrow Fe(OH)_3(s)$$

Non-Metal Anions

- Carbonates and sulfates are anions that can be identified by adding a few drops of barium chloride solution followed by dilute hydrochloric acid.
- Barium cations react with the anion to form a white precipitate.
- When hydrochloric acid is added:
  - if the white precipitate remains, a sulfate anion is present
  - if carbon dioxide is given off, a carbonate anion is present.

barium chloride solution + sodium sulfate solution ⟶
  sodium chloride solution + solid barium sulfate

HT Carbonates: $Ba^{2+}(aq) + CO_3^{2-}(aq) \longrightarrow BaCO_3(s)$
Sulfates: $Ba^{2+}(aq) + SO_4^{2-}(aq) \longrightarrow BaSO_4(s)$

> **Key Point**
>
> Precipitates may take a while to form. You need to check the reaction after a few minutes, otherwise the precipitate may be missed.

The white precipitate disappears when hydrochloric acid is added. Carbon dioxide is given off.

The white precipitate remains when hydrochloric acid is added.

Chloride, Bromide and Iodide Anions

- To test for the presence of **halide** anions, dilute nitric acid is added followed by silver nitrate.
- When the silver ions react with a halide ion a coloured precipitate is formed:
  - Chlorides form a white precipitate.
  - Bromides form a cream precipitate.
  - Iodides form a pale yellow precipitate.

$Ag^+(aq) + Cl^-(aq) \rightarrow AgCl(s)$

$Ag^+(aq) + Br^-(aq) \rightarrow AgBr(s)$

$Ag^+(aq) + I^-(aq) \rightarrow AgI(s)$

> **Quick Test**
>
> 1. Describe the test for hydrogen gas.
> 2. What is the test for ammonium (NH_4^+) ions?
> 3. A solution of iron ions is reacted with sodium hydroxide and makes a green precipitate. Which iron ion must have been present?

> **Key Words**
>
> displacement
> reactivity series
> precipitate
> halide

Ion Tests and Instrumental Methods of Analysis

You must be able to:

- Describe how to carry out a flame test and analyse the results
- Describe the advantages of instrumental methods of analysis
- Interpret the results of instrumental analysis when given reference data.

Flame Tests

- Each element on the periodic table emits a characteristic **spectrum** of light when heated at a high temperature.
- Metal cations tend to produce very bright, recognisable colours when heated.
- A flame test can, therefore, be used to identify the metal cation present:
  1. Dip a clean **nichrome** wire into concentrated hydrochloric acid.
  2. Dip the wire into the solid that is being tested.
  3. Use the wire to place the sample into the hottest part of the Bunsen burner flame.
  4. Note the colour that is produced.
- Hydrochloric acid is used so that some of the solid is converted into a metal chloride.
- Metal chlorides are more volatile than other salts, so the flame test is more likely to work.
- Flame tests are **qualitative** – a person must identify the colour that is produced.
- Other observers may perceive differences in the colour.
- Flame tests are effective if there is only one metal cation present.
- If there is more cation the stronger colour may mask the others.
- Elements in Group 1 do not form precipitates with hydroxide ions, so flame testing is the only way to identify a Group 1 cation.

Instrumental Methods of Analysis

- Qualitative observations can vary according to the observer.
- It is more reliable to remove the judgement of the observer and use an instrument to measure the results.
- There are instruments that have sensors, which can detect a substance.
- The instrument then relays the result in different ways:
  - as a reading (e.g. pH probe)
  - as a graph (e.g. vehicle emissions)
  - as a chart or table (e.g. concentrations of different ions in a soil sample).
- **Instrumental analysis** has the following advantages:
  - It is **quantitative**.
  - It is more sensitive (can detect and report far lower quantities).
  - It is more accurate.
  - It is able to measure factors more rapidly.

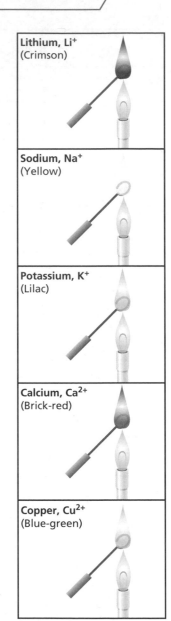

Lithium, Li⁺
(Crimson)

Sodium, Na⁺
(Yellow)

Potassium, K⁺
(Lilac)

Calcium, Ca²⁺
(Brick-red)

Copper, Cu²⁺
(Blue-green)

Key Point

The flame colours for common cations are:

Li⁺ = crimson

Na⁺ = yellow

K⁺ = lilac

Ca²⁺ = brick-red

Cu²⁺ = blue-green

- **Spectrometers** can detect and measure the light produced in a flame test to give a much more accurate result.
- They detect the different light wavelengths emitted when elements are burned.

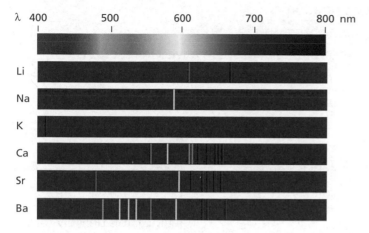

- The results of an instrumental analysis can be easily interpreted when a set of **reference data** is provided.

A number of cars had a vehicle emission test:

MOT			
Vehicle Emission Test	Car 1	Car 2	Car 3
Component	g/km	g/km	g/km
Hydrocarbons	1.6	2.6	1.74
Carbon monoxide	12	15	12.06
NO_x	0.67	0.86	1.22
CO_2	230	258	300

Reference data for MOT vehicle emissions:

Component	g/km	
Hydrocarbons	<1.75	
Carbon monoxide	<13.06	Levels higher = Fail
NO_x	<0.87	
CO_2	<258	

Car 1 passes the emission test.
Car 2 fails (the hydrocarbon and carbon monoxide readings are higher than the given levels).
Car 3 fails (the NO_x and CO_2 readings are higher than the given levels).

Quick Test

1. In a flame test, a lilac flame is observed. What cation is present?
2. What colour flame is produced when Li⁺ ions are present?
3. Give **two** advantages of instrumental methods of analysis.

Monitoring Chemical Reactions

You must be able to:

- Describe how to carry out a titration
- HT Explain what is meant by concentration of a solution
- HT Describe how to calculate the volume of gases involved in reactions.

Titration

- **Titration** is the method used to find out how much acid is needed to neutralise an alkali or vice versa:
  1. Measure the alkali into a conical flask using a pipette.
  2. Add a few drops of indicator to the conical flask.
  3. Fill the burette with acid, recording the start volume.
  4. Add acid slowly to the alkali until the indicator just changes colour (the end point).
  5. Record the end volume of acid.
  6. Work out how much acid has been added. This is called the **titre** (titre = final volume – start volume).
  7. Repeat the titration until you have a set of **concordant** (similar) results. Then calculate an average (mean).

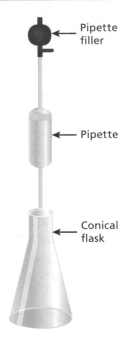

Pipette filler

Pipette

Conical flask

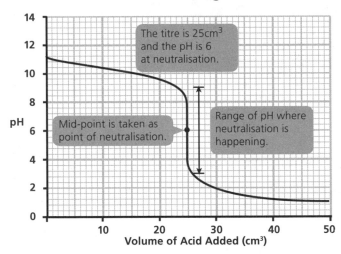

The titre is 25cm³ and the pH is 6 at neutralisation.

Mid-point is taken as point of neutralisation.

Range of pH where neutralisation is happening.

pH / Volume of Acid Added (cm³)

- pH curves can be used to show the pH changes in a titration.
- You should be able to read and interpret pH curves to work out:
  - the titre (the volume of acid needed to neutralise the alkali)
  - the pH when a certain amount of acid has been added.

HT Concentration of Solutions

- **Concentration** is a measure of how many moles of a compound (the solute) are dissolved in 1dm³ of solvent.
- The number of moles is based on the relative molecular mass.
- For example, sodium chloride, NaCl, has the relative formula mass of 58.5g.
- If 58.5g of NaCl is added to 1dm³ of H_2O, there is 1 mole present in solution – it has a concentration of 1.0mol/dm³.

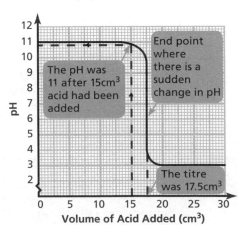

The pH was 11 after 15cm³ acid had been added

End point where there is a sudden change in pH

The titre was 17.5cm³

pH / Volume of Acid Added (cm³)

Volume and Concentration

- Titrations work because there is a relationship between the volume of one solution and the concentration of another.
- The burette contains a solution of unknown concentration.
- The flask contains the **standard solution**, i.e. a solution of precise volume and known concentration.
- Once the unknown solution has been neutralised, the titre used to neutralise it is recorded.
  1. The amount of moles in the titre = concentration × volume.
  2. The amount in moles of the unknown solution is worked out (using stoichiometry).
  3. The concentration of the unknown solution is calculated:

$$\text{concentration} = \frac{\text{moles}}{\text{volume}}$$

> **Key Point**
>
> Drawing a pH curve requires accuracy. Use a sharp pencil and ensure all axes are labelled.

In a titration, $0.025\,dm^3$ of $0.2\,mol/dm^3$ hydrochloric acid, HCl, is neutralised by $0.0175\,dm^3$ of sodium hydroxide solution, NaOH.

$$NaOH + HCl \rightarrow NaCl + H_2O$$

What is the concentration of the NaOH solution?

moles of HCl used = concentration × volume
$$= 0.2\,mol/dm^3 \times 0.025\,dm^3 = 0.005\,mol$$

concentration NaOH(aq) $= \dfrac{0.005\,mol}{0.0175\,dm^3} = 0.29\,mol/dm^3$
(to 2 significant figures)

> The ratio of NaOH to HCl is 1 : 1, so number of moles of NaOH must also be 0.005mol.

Molar Amounts of Gases and their Volumes

- One mole (1mol) of all gases will occupy the same volume (at the same temperature and pressure).
- This volume is called the **molar volume** and (under room temperature and pressure) it is $24\,dm^3$.
- The relationship between the volume and number of moles of gas is given by the equation:

$$\text{moles of gas} = \frac{\text{volume of gas}}{\text{molar volume}}$$

> If the amount of gas in moles is known, then the volume of gas can be calculated by rearranging the equation.

 Quick Test

1. How is the titre calculated?
2. What is the equation that can be used to calculate the concentration of an unknown solution?
3. **HT** Define the molar volume of a gas.

> **Key Words**
>
> titration
> titre
> concordant
> **HT** concentration
> **HT** standard solution
> **HT** molar volume

Calculating Yields and Atom Economy

You must be able to:

- Calculate theoretical and actual yields of a reaction
- Work out the atom economy for a given reaction
- Explain why a particular reaction pathway is chosen to produce a specified product.

Calculating Theoretical Yield

- The balanced equation for a chemical reaction can be used to calculate the **theoretical yield** of the products.
- This is achieved by calculating the molar masses of the different reactants and products.

> **Key Point**
>
> Always show the working for your calculations.

$CuO(s) + H_2SO_4(aq) \longrightarrow CuSO_4(aq) + H_2O(l)$

How much $CuSO_4(aq)$ can be produced from 39.75g of CuO(s)?

molar mass of CuO(s) = 63.5 + 16.0 = 79.5
molar mass of $CuSO_4(aq)$ = 63.5 + 32.1 + (4 × 16) = 159.6

79.5g of CuO(s) produces 159.6g of $CuSO_4(aq)$ ←
39.75g of CuO(s) produces 79.8g of $CuSO_4(aq)$ ←

> 39.75g of CuO(s) is half this amount, so half of 159.6g $CuSO_4(aq)$ will be produced.

> This is the theoretical yield.

Calculating Percentage Yield

- In reality, sometimes the amount of product formed may be less than expected. For example:
  - Some reactant may be unable to react if it is covered by a layer of newly formed product.
  - Some product may escape from the reaction vessel.
  - Some reactions are reversible and the products may react to re-form reactants.
- The **actual yield** is the amount of product actually produced by a reaction.
- The **percentage yield** is calculated using the following equation:

$$\text{percentage yield (\%)} = \frac{\text{actual yield}}{\text{theoretical yield}} \times 100$$

- A 100% yield means that all of the reactants successfully reacted to form the products.

The reversible reaction between carbon dioxide and hydrogen makes methane and water.
A student predicts that 180g water will be produced using this reaction in an experiment. When the experiment is carried out, the actual yield of water is 80g.

Calculate the percentage yield of water.

$$\text{percentage yield (\%)} = \frac{80}{180} \times 100 = 44.4\%$$

Atom Economy

- **Atom economy** is a measure of how many of the atoms in the reactants are converted into the desired product of a reaction.
- An atom economy of 100% means that every atom in the reactants is used to produce the desired product.
- In reality, not all atoms in the reactants will go into the desired product. Some may form other products (by-products) instead.
- The higher the atom economy, the more **efficient** the reaction.

$$\text{atom economy (\%)} = \frac{\text{relative formula mass of the desired product}}{\text{sum of the relative formula masses of all the reactants}} \times 100$$

Work out the atom economy of this reaction to make water, H_2O:

$$CuO(s) + H_2SO_4(aq) \rightarrow CuSO_4(aq) + H_2O(l)$$

relative formula mass of $H_2O = (2 \times 1) + 16 = 18$

sum of relative formula masses of reactants $= 79.5 + 98.1 = 177.6$

atom economy $= \dfrac{18}{177.6} \times 100 = 10.1\%$

> relative formula mass of CuO = $63.5 + 16 = 79.5$

> relative formula mass of $H_2SO_4 =$ $(2 \times 1) + 32.1 + (4 \times 16) = 98.1$

> This reaction has a low atom economy – it would not be a good reaction to use if you wanted to produce water.

HT Choosing a Reaction Pathway

- There are often many different chemical reactions that could be used to produce the same product.
- In industry, chemists will decide on which **reaction pathway** to follow, based upon a number of factors:
  1. Atom economy – how efficient is the reaction at producing the desired product?
  2. Environmental impact – how environmentally friendly is the reaction / will it produce a lot of hazardous waste?
  3. Usefulness of by-products – can the by-products be used (rather than being wasted)?
  4. Cost – how expensive is the reaction? If an efficient reaction is too expensive, a less costly but possibly less efficient pathway may be chosen.
  5. Type of reaction – the reaction may be reversible and, therefore, optimum conditions would have to be used to prevent it reacting in the wrong direction.
  6. Speed – how fast is the reaction? A fast reaction is preferable, but if it is too fast there is the danger of an explosion.
  7. Yield/Equilibrium position – to what extent are all of the reactants converted into the products? Some reactions are reversible (see page 60) which means that a mixture of reactants and products are formed. The yield or equilibrium position tells you the balance of products to reactants.

> **Key Point**
>
> Scientists look at the evidence and then present reasoned explanations for choosing one pathway over others. They might have to present the evidence in a variety of ways, both on paper and in electronic forms.

> **Key Words**
>
> theoretical yield
> actual yield
> percentage yield
> atom economy
> efficient
> HT reaction pathway

Quick Test

1. 160 tonnes of CuO are reacted with H_2SO_4.
   a) What is the theoretical yield of $CuSO_4$?
   b) What is the atom economy of the reaction?

Controlling Chemical Reactions

You must be able to:

- Suggest methods for working out the rate of reaction
- Describe factors that affect the rate of reaction
- Explain how surface area to volume ratios change in solids.

Rates of Reaction

- The **rate of reaction** is a measure of how much product is made in a specific time.
- It can be measured in g/s or g/min for mass changes or cm^3/s or cm^3/min for volume changes.
- Chemical reactions stop when one of the reactants is used up.
- If a gas is produced during an experiment then the rate of reaction can be determined by carrying out the reaction on a balance and timing how long it takes/recording the mass at regular intervals until there is no further change.
- The rate of reaction can also be determined by collecting the gas produced e.g. in a gas syringe and timing how long it takes to collect a certain volume of gas/recording the volume of gas collected at regular intervals until there is no further change.
- Some reactions involve a colour change or the formation of a solid that turns the reaction mixture opaque. A common way of measuring the rate of these reactions is to time how long it takes for the reaction to no longer be transparent.
- The amount of product produced depends on the amount of reactant(s) used.
- Often there is an excess of one of the reactants.
- The one that is used up first is the limiting reactant.

Changing the Temperature

- Chemical reactions happen when particles collide with enough energy for the reactants to bond together or split apart.
- The more successful **collisions** there are between particles, the faster the reaction.
- At low temperatures, the particles move slowly as they have less **kinetic energy**. This means that the particles collide less often, and with lower energy, so fewer collisions will be successful – the rate of reaction will be slow.
- At high temperatures, the particles move faster as they have more kinetic energy. This means that particles successfully collide more often, and with higher energy, giving a faster rate of reaction.

Low Temperature | High Temperature

With a Catalyst

Manganese(IV) oxide (catalyst)

Changing the Concentration of Liquids

- In a reaction where one or both reactants are in low concentrations, the particles will be spread out in solution.
- The particles will collide with each other less often, so there will be fewer successful collisions.
- If there is a high concentration of one or both reactants, the particles will be crowded close together.
- The particles will collide with each other more often, so there will be many more successful collisions.
- To measure the effect of concentration on rate of reaction, create a series of different concentrations of a reactant by diluting a stock solution of known concentration.
- Once a range of at least five different concentrations has been created, the reactants can be reacted together.

Low Concentration

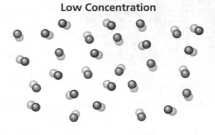

High Concentration

Changing the Pressure of Gases

- When a gas is under a low **pressure**, the particles are spread out. The particles will collide with each other less often, so there will be fewer successful collisions.
- When the pressure is high, the particles are crowded more closely together. The particles therefore collide more often, resulting in many more successful collisions.
- It difficult to carry out an experiment where gas pressure is changed in a school laboratory. Specialised equipment is needed to ensure that the pressure can be maintained and measured.

Low Pressure

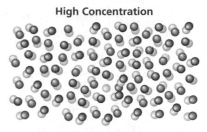

High Pressure

Changing the Surface Area of Solids

- The larger the **surface area** of a solid reactant, the faster the reaction.
- Powdered solids have a large surface area compared to their volume – they have a high surface area to volume ratio.
- This means there are more particles exposed on the surface for the other reactants to collide with.
- The greater the number of particles exposed, the greater the chance of them colliding successfully, which increases the rate of the reaction.
- As a result, powders react much faster than a lump of the same reactant.
- To measure the effect of surface area on rate of reaction, take the reactant and create a set of different surface areas, e.g. by cutting the solid into progressively smaller sections.
- Once a suitable range of surface areas has been created, they can then be reacted with the other reactant.

| Lump of Solid (Large Particles) | Powdered Solid (Small Particles) |

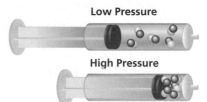
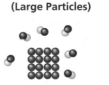

> ### Quick Test
>
> 1. What is the effect of increasing the temperature of a reaction mixture?
> 2. How could the rate of reaction be increased in a reaction involving gaseous reactants?
> 3. Describe the surface area to volume ratio of a large block of calcium carbonate compared to calcium carbonate powder.

Key Words

rate of reaction
collisions
kinetic energy
pressure
surface area

Catalysts and Activation Energy

You must be able to:

- Analyse the rate of reaction using a graph
- Describe the characteristics of catalysts and their effect on the rate of reaction
- Explain the action of catalysts in terms of activation energy.

Analysing the Rate of Reaction

- From a graph you can find out the following:
  1. How long it takes to make the maximum amount of products:
     Draw a vertical line from the beginning of the flat line (which indicates the reaction has finished) down to the x-axis (time).
  2. How much product was made:
     Draw a horizontal line from the highest point on the graph across to the y-axis.
  3. Which reaction is the quicker:
     Compare the gradient (steepness) of graphs for the same reaction under different conditions.

- $\frac{1}{time}$ is **proportional** to the **rate of reaction**.

Effect of Catalysts

- A **catalyst** is a substance that speeds up the rate of a chemical reaction without being used up or changed in the reaction.
- Catalysts are very useful materials, as only a small amount is needed to speed up a reaction involving large amounts of reactant.
- You can see how a catalyst affects the rate of reaction by comparing graphs of reactions with and without the catalyst.
- The graph on the right shows two reactions that eventually produce the same amount of product.
- One reaction takes place much faster than the other because a catalyst is used.
- As catalysts are not used up in a reaction, they are not a reactant.
- When a catalyst is used in a reaction, the name of the catalyst is written above the arrow in the equation.
- For example, hydrogen peroxide is relatively stable at room temperature and pressure. It breaks down into water and oxygen very slowly.
- In the presence of the **inorganic** catalyst, manganese(IV) oxide, the breakdown is very rapid.

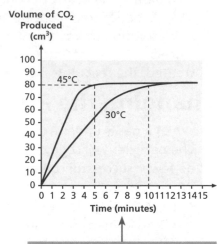

Volume of CO_2 Produced (cm^3)

The line is steeper for the 45°C line, so the rate is faster.

relative rate of reaction for
$45°C = \frac{1}{5min} = 0.2$

relative rate of reaction for
$30°C = \frac{1}{10min} = 0.1$

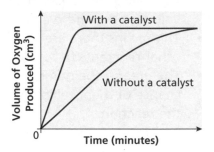

Key Point

When looking at a graph, it may be obvious which line is steeper but it pays to calculate the gradient as well.

Without a catalyst: hydrogen peroxide $\longrightarrow$ water + oxygen (very slow)

With a catalyst: hydrogen peroxide $\xrightarrow{\textit{manganese(IV) oxide}}$ water + oxygen (very fast)

Catalysts and Activation Energy

- Catalysts work by lowering the **activation energy** for a reaction.

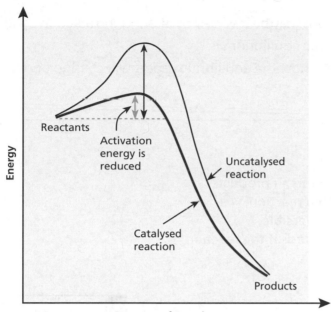

Progress of Reaction

- This means that the reaction is more likely to happen, because less energy is needed to get it started.
- This may be achieved by providing a surface for the reactants to temporarily **adhere** to (stick to).
- Adhering to the catalyst's surface uses electrons, weakening the bonds in the reactants, lowering the activation energy.

Surface of catalyst

Biological Catalysts

- **Enzymes** and certain pigments are biological catalysts.
- They are proteins and, as with inorganic catalysts, they lower the activation energy for a reaction.
- Examples include:
  - **amylase** – an enzyme that breaks down the starch (a polymer) into glucose
  - **chlorophyll** – a pigment that catalyses photosynthesis
  - **catalase** – an enzyme that, like manganese(IV) oxide, can break down hydrogen peroxide into water and oxygen.

Key Point

Don't forget, catalysts are not used up in a reaction.

Key Words

proportional
rate of reaction
catalyst
inorganic
activation energy
adhere
enzymes

Quick Test

1. What does the gradient (steepness) of a reaction graph tell you?
2. How do catalysts affect activation energy?
3. Which biological molecules can catalyse a reaction?

Equilibria

You must be able to:

- Recall that some reactions may be reversed by altering the reaction conditions
- Describe what a dynamic equilibrium is
- **HT** Make predictions about how the equilibrium point will change under different conditions.

Reversible Reactions

- Most chemical reactions are of the type: reactants → products
- The reaction is complete, i.e. it is an **irreversible** reaction.
- There are, however, some reactions that are **reversible**.
- A reversible reaction can go forwards or backwards if the reaction conditions are changed.

Dynamic Equilibrium

- A reversible reaction can reach **equilibrium** (a balance).
- This means that the rate of the forward reaction is equal to the rate of the reverse reaction.
- A closed system is one where the conditions of the reaction (such as pressure and temperature) are not changed and no substances are added or removed.
- At equilibrium in a closed system:
  - there is no change in the amounts and concentrations of reactants and products
  - reactants are reacting to form products and products are reacting to re-form reactants at the same rate.
- This is called a **dynamic equilibrium**.
- For example, heating a mixture of hydrogen and iodine gas:

> **Heating a mixture of hydrogen and iodine gas.**
> $$H_2(g) \; + \; I_2(g) \; \rightleftharpoons \; 2HI(g)$$

  - Unlike an irreversible reaction, the reactants will *not* completely form hydrogen iodide.
  - When the reaction starts, the forward reaction producing HI will dominate.
  - This is because the concentration of HI will be very low.
  - As the concentration of HI increases, the rate of the reverse reaction (making H_2 and I_2) will also increase.
  - Eventually the rate of the forward reaction will equal the rate of the reverse reaction.
  - The reaction is in dynamic equilibrium.

HT Changing the Equilibrium

- The position of the equilibrium can be changed by altering:
  - the temperature
  - the pressure
  - the concentration of the reactant(s) and/or the product(s).

> **Key Point**
>
> The symbol indicates a reversible reaction.

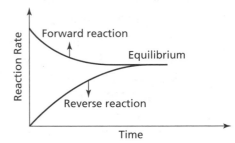

> **Key Point**
>
> Remember, at equilibrium in a closed system the concentrations of the reactants and products remain the same.

- If the equilibrium lies to the right of the reaction, the concentration of the products is greater than the concentration of the reactants.
- If the equilibrium lies to the left of the reaction, the concentration of the products is less than the concentration of the reactants.

HT Le Chatelier's Principle

- **Le Chatelier's principle** states that when the conditions of a system are altered, the position of the equilibrium changes to try and restore the original conditions.
- This means that predictions can be made about:
  - what will happen when certain conditions are changed
  - what conditions need to be changed to produce the highest yield of a substance.

HT Changing Conditions

- If a reaction is **exothermic** in the reactant → product direction, Le Chatelier's principle indicates that increasing the temperature will:
  - decrease the yield of products
  - increase the amount of reactants (to restore the original conditions).
- If a reaction is **endothermic** in the reactant → product direction, then an increase in temperature will:
  - increase the yield of products
  - decrease the amount of reactants (to restore the original conditions).
- Pressure changes only apply to reactions involving gases.
- Pressure changes will alter the equilibrium if there is a difference in the amount of gas in the products and the reactants.
- Applying Le Chatelier's principle, if there are more moles of gas reactants compared to products:
  - an increase in pressure will lead to more product
  - a decrease in pressure will cause the equilibrium to move towards the side of the equation with the greatest number of moles.
- If the concentration of one of the substances in a reaction increases, then Le Chatelier's principle means that the equilibrium will shift to restore the original conditions:
  - If the concentration of a reactant is increased, more product will be made.
  - If the concentration of a product is increased, more reactant will be made.

Key Point

With reversible reactions, you have to be clear which direction is exothermic. If the forward reaction is exothermic, the reverse reaction will be endothermic and vice versa.

Key Point

A catalyst does not affect the equilibrium position. It speeds up the rate of reaction, so equilibrium will be reached more rapidly.

Key Words

irreversible
reversible
equilibrium
dynamic equilibrium
HT Le Chatelier's principle
HT exothermic
HT endothermic

Quick Test

1. HT Name **three** factors that can change the position of the equilibrium for a reaction.
2. No reactions take place when a reversible reaction is at equilibrium. True or false?
3. HT If a forward reaction is exothermic, what would the effect of increasing the temperature be? Explain your answer.

Review Questions

Introducing Chemical Reactions

1 Write down the number of atoms of each element in each of the following compounds.

a) H_2SO_4 [1]

b) $Cu(NO_3)_2$ [1]

c) CH_3CH_2COOH [1]

d) C_2H_6 [1]

2 Balance the following equations:

a) $CuO(s) + H_2SO_4(aq) \rightarrow CuSO_4(aq) + H_2O(l)$ [2]

b) $Mg(s) + O_2(g) \rightarrow MgO(s)$ [2]

c) $Mg(OH)_2(aq) + HCl(aq) \rightarrow MgCl_2(aq) + H_2O(l)$ [2]

d) $CH_4(g) + O_2(g) \rightarrow CO_2(g) + H_2O(l)$ [2]

3 The table below shows the names and formulae of some common ions.

Fill in the missing information to complete the table.

Name of Ion	Formula
Carbonate	
Lithium	
	Fe^{3+}
	O^{2-}
Sulfate	

[5]

4 HT Write the half equation for each of the following reactions:

a) Solid lead to lead ions [1]

b) Aluminium ions to aluminium [1]

c) Bromine to bromide ions [1]

d) Silver ions to solid silver [1]

Total Marks .... / 21

Chemical Equations

1 HT When writing a balanced ionic equation, which species appear in the equation? [1]

2 HT Write the net ionic equation for:

HT **a)** $AgNO_3(aq) + KCl(aq) \rightarrow AgCl(s) + KNO_3(aq)$ [1]

b) magnesium nitrate (aq) + sodium carbonate (aq) $\rightarrow$

magnesium carbonate (s) + sodium nitrate (aq) [1]

Total Marks _____ / 3

Moles and Mass

1 HT Calculate the **number of moles** of each of the following elements:

a) 6.9g of Li [1]

b) 62g of P [1]

2 HT Calculate the molar mass of ammonium chloride, NH_4Cl.
(The relative atomic mass of H = 1, Cl = 35.5 and N = 14.) [1]

3 HT Calculate the mass of one atom of each of the following elements.
Use the periodic table on page 140 to help you.

a) Tungsten [1]

b) Tin [1]

4 HT Barium chloride reacts with magnesium sulfate to produce barium sulfate and magnesium chloride.

What mass of barium sulfate will be produced if 5mol of barium chloride completely reacts?
Show your working. [2]

5 HT How many moles are there in 22g of butanoic acid, $C_4H_8O_2$?

    A   0.1   **C**   0.5

    B   0.25   **D**   1   [1]

Total Marks _____ / 8

Energetics

1   Atu pulls a muscle whilst playing rugby.
A cold pack is applied to his leg to help cool the muscle and prevent further injury.
The pack contains ammonium nitrate and water.
When the pack is crushed, the two chemicals mix and ammonium nitrate
dissolves endothermically.

    a)   What is meant by the term **endothermic**?   [1]

    b)   Where does the energy come from that enables the pack to work?   [1]

2   HT Mark reacts hydrogen gas with chlorine gas:

    $H_2(g) + Cl_2(g) \rightarrow 2HCl(g)$

Bond	Bond Energies (kJ/mol)
H–Cl	431
H–H	436
Cl–Cl	243

    a)   Calculate the energy change for the reaction and state whether the reaction is
         endothermic or **exothermic**.   [3]

    b)   Draw the expected reaction profile for the reaction.   [1]

Total Marks _____ / 6

Types of Chemical Reactions

1 Which **two** of the following reactions are oxidation reactions?

    A aluminium + oxygen → aluminium oxide

    B sodium chloride + silver nitrate → silver chloride + sodium nitrate

    C copper sulfate + sodium hydroxide → copper hydroxide + sodium sulfate

    D copper + oxygen → copper oxide   [2]

2 Look at the following reaction:

iron(III) oxide + carbon monoxide → iron + carbon dioxide

    a) Which species is being **reduced**?   [1]

    b) Which species is being **oxidised**?   [1]

    c) Write the balanced symbol equation for the reaction.   [2]

3 HT Claudia places a copper wire into a solution of colourless silver nitrate solution.

Time ⟶

As time passes, Claudia notices that shiny crystals start developing on the surface of the copper wire.
She also notices that the solution becomes a light blue colour.

    a) Write the balanced symbol equation for the reaction between copper and silver nitrate.   [2]

    b) What are the shiny crystals on the wire?   [1]

c) What causes the blue coloration of the solution? [1]

d) Which chemical species are being oxidised and which are being reduced?
You must explain your answer. [2]

Total Marks _____ / 12

pH, Acids and Neutralisation

1 Underline the **acid** in each reaction.

a) $Mg(OH)_2(aq) + 2HCl(aq) \rightarrow MgCl_2(aq) + 2H_2O(l)$ [1]

b) $H_2SO_4(aq) + 2NaOH(aq) \rightarrow Na_2SO_4(aq) + 2H_2O(l)$ [1]

c) $2CH_3COOH(aq) + 2Na(s) \rightarrow H_2(g) + 2CH_3COONa(aq)$ [1]

d) $2HF(aq) + Mg(s) \rightarrow MgF_2(aq) + H_2(g)$ [1]

2 Part of the reactivity series is shown in the diagram on the right.
When a metal is reacted with an acid it forms a metal salt, plus hydrogen
gas. For example:

lead + sulfuric acid → lead sulfate + hydrogen

calcium + sulfuric acid → calcium sulfate + hydrogen

Reactivity Series

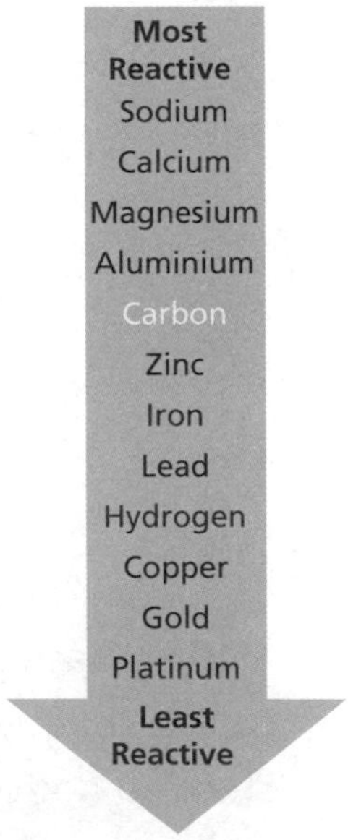

Most Reactive
Sodium
Calcium
Magnesium
Aluminium
Carbon
Zinc
Iron
Lead
Hydrogen
Copper
Gold
Platinum
Least Reactive

a) Which of the two reactions has the fastest initial reaction? [1]

b) A metal **X** is reacted with sulfuric acid.
It reacts violently compared with the other two reactions.

Where would **X** be placed on the reactivity series? [1]

3 Nitric acid and sodium hydroxide are reacted together.

a) Write the balanced symbol equation for the reaction. [2]

b) Which ions are spectator ions? [2]

c) Rewrite the equation you wrote for part **a)** showing only the reacting species. [2]

4 HT What is meant by the term **strong acid**? [1]

5 How many more times concentrated are the H^+ ions in a solution with a pH of
6 compared to a solution with a pH of 2? [1]

Total Marks _____ / 14

Electrolysis

1 In what state(s) will ionic compounds conduct electricity? [1]

2 Masum is carrying out the electrolysis of water and sulfuric acid.

a) Which of the following would be the most appropriate
material for the electrodes?

A Wood

B Copper

C Carbon

D Plastic [1]

b) Write the names of the anions and cations involved in this electrolysis. [2]

c) Write the reactions taking place at **i)** the anode and **ii)** the cathode. [2]

3 Electrolysis is used to copper-plate objects.

The diagram below shows the apparatus for electroplating
using copper and a metal object in copper(II) sulfate solution.

What are ions **X** and **Y**? [2]

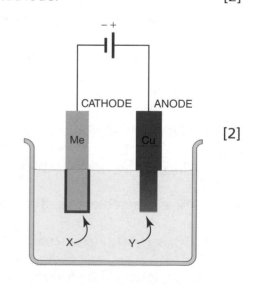

Total Marks _____ / 8

Predicting Chemical Reactions

1. Which of the following statements is **correct** about a Group 7 element?

   A It conducts electricity.

   B It is malleable.

   C It dissolves in water to make bleach.

   D It reacts with water to form hydrogen.   [1]

2. How many electrons are in the outermost shell of a Group 1 element?   [1]

3. Sort the following elements into order of reactivity, with the most reactive first:

   bromine   chlorine   fluorine   iodine   [1]

4. a) Draw the electronic structure of sodium (atomic number 11) and potassium
      (atomic number 19).   [2]

   b) Sodium is placed into a container of water.
      It moves around rapidly on top of the water and burns with a yellow flame.

      Describe what happens when potassium is placed into a container of water. [2]

   c) Which of the following best describes why potassium reacts differently to sodium?

      A Sodium is denser than potassium.

      B Potassium is denser than sodium.

      C Potassium's outermost electron is closer to the nucleus.

      D Potassium's outermost electron is further away from the nucleus.   [1]

5. Look at the periodic table on page 140.

   Which is the **least** reactive Group 2 metal?   [1]

6. Zinc is reacted with sulfuric acid:

   $Zn(s) + H_2SO_4(aq) \rightarrow ZnSO_4(aq) + H_2(g)$

   What **ions** will be formed in this reaction?   [1]

   Total Marks _____ / 10

Identifying the Products of Chemical Reactions

1 Chloe reacts an unknown with hydrochloric acid.
A gas is formed, which Chloe tests using limewater.
The limewater turns a milky white colour.

What gas has been produced?

A Hydrogen   **C** Carbon dioxide

B Oxygen   **D** Methane   [1]

2 Describe the test for hydrogen gas.   [2]

3 Harry wants to identify a mystery metal ion.

Ion	Precipitate Colour
Calcium hydroxide	white
Copper hydroxide	light blue
Iron(II) hydroxide	green
Iron(III) hydroxide	red / brown

Harry mixes a solution containing the mystery metal ion with sodium hydroxide.
He observes that a green precipitate is formed.

a) What cation is in the mystery solution?   [1]

b) Harry learns that the mystery solution contains sulfate ions.

What is the formula of the ionic compound?   [1]

4 **HT** Copper(II) nitrate is reacted with sodium hydroxide.

Write the ions that react in this reaction.   [2]

5 Ruth adds barium chloride to a mystery solution.
A white precipitate forms.
The precipitate remains after adding hydrochloric acid.

Write the formula for the precipitate.   [1]

Total Marks _____ / 8

Practice Questions

Ion Tests and Instrumental Methods of Analysis

1 A mystery acid is tested by adding dilute nitric acid followed by silver nitrate solution. A white precipitate forms.

Which ion has been identified?

A Silver

B Nitrate

C Chloride

D Sulfate [1]

2 What happens to the pH of an acidic solution when an alkali is gradually added? [1]

3 HT Magnesium sulfate has a relative molecular mass of 120.

a) Calculate the mass of magnesium sulfate needed to make 1dm³ of 0.05mol/dm³ solution. [3] Show your working.

b) What would happen to the concentration of the solution if the amount of magnesium sulfate was kept the same but the final volume of solution was halved? [1]

4 Forensic scientists are testing a car to see if it was involved in a hit and run accident on another car. A sample of paint is taken from the car and tested using a flame test.

a) If the paint sample only contained potassium ions, what colour flame would be seen?

A Crimson

B Lilac

C Yellow-red

D Green-blue [1]

b) The paint sample actually contains a mixture of metal ions.

Why does this make the results difficult to interpret? [1]

5 MOT testing stations use instruments that record the level of pollutants in vehicle emissions.

Give **two** reasons why instrumental analysis is an effective way of measuring pollutant levels. [2]

Total Marks _____ / 10

Monitoring Chemical Reactions

1 Maara carries out a titration of sodium hydroxide and sulfuric acid. He sets up a titration using the equipment shown on the right.

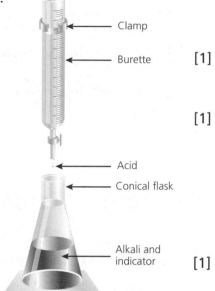

a) Explain why Maara uses an indicator. **[1]**

b) What safety precautions should Maara take when carrying out the experiment? **[1]**

c) Maara slowly adds the acid from the burette until the sodium hydroxide is just neutralised.
He then measures the volume of the sulfuric acid again.

Describe how Maara can tell when the sodium hydroxide solution is just neutralised. **[1]**

2 The diagram on the right shows a burette before and after a titration.

What is the correct reading for the titre?

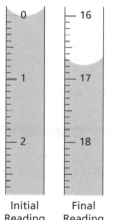

A 16.1cm³

B 16.8cm³

C 17.2cm³

D 0.9cm³ **[1]**

Initial Reading   Final Reading

3 Caitlin carried out a titration of sodium hydroxide solution and hydrochloric acid:

$$NaOH(aq) + HCl(aq) \rightarrow NaCl(aq) + H_2O(l)$$

She found that 0.05dm³ of 0.2mol/dm³ HCl was neutralised by 0.035dm³ of NaOH solution.

Calculate the concentration of the NaOH solution.
Give your answer to two significant figures. **[3]**

4 Nicki has 2.00mol of hydrogen gas at room temperature and pressure.
The molar volume is 24dm³.

a) Calculate the volume of gas. **[1]**

b) Describe what would happen if the gas was heated. **[2]**

Practice Questions

5 Which of the following is the correct equation for calculating the molar volume of a gas?

A amount (moles) of gas = $\dfrac{\text{volume of gas}}{\text{molar volume of gas}}$

B molar volume = $\dfrac{\text{volume of gas}}{\text{amount (moles) of gas}}$

C molar volume = volume of gas × amount (moles) of gas

D molar volume = $\dfrac{\text{amount (moles) of gas}}{\text{volume of gas}}$ [1]

Total Marks _____ / 11

Calculating Yields and Atom Economy

1 Sanjay plans to react magnesium oxide with sulfuric acid:

$$MgO(s) + H_2SO_4(aq) \rightarrow MgSO_4(aq) + H_2O(l)$$

He uses 40.3g of MgO.

a) How much $MgSO_4$ should he theoretically make? [1]

b) At the end of the experiment, Sanjay records that 100g of $MgSO_4$ had been made.

Calculate the percentage yield for the reaction. [1]

2 Which of the reactions below is more efficient at generating hydrogen?

A $C(s) + 2H_2O(g) \rightarrow CO_2(g) + 2H_2(g)$

B $2Li(s) + H_2SO_4(aq) \rightarrow Li_2SO_4(aq) + H_2(g)$ [1]

3 HT Different chemical reactions can be used to make the same product.

Give **three** factors that chemists consider when choosing which reaction pathway to follow. Explain how the factors you have chosen will affect the outcome. [3]

Total Marks _____ / 6

Controlling Chemical Reactions

1 Niamh is investigating the effect of temperature on the breakdown of hydrogen peroxide in the presence of the catalyst manganese(IV) oxide.

In each experiment Niamh uses a different temperature. All other factors are kept the same.

Niamh's results are shown in the graph below.

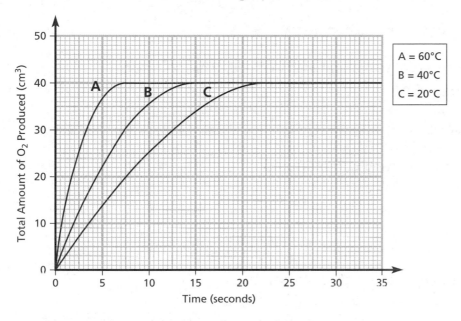

A = 60°C
B = 40°C
C = 20°C

a) In which experiment is the reaction the **slowest**? [1]

b) Write a suitable conclusion for the investigation. [1]

2 Which of the following could affect the rate of reaction?

A Time

B Surface area

C Concentration

D Volume [3]

Total Marks _____ / 5

Practice Questions

Catalysts and Activation Energy

1 Explain, in terms of activation energy, how a catalyst affects the rate of reaction. [2]

2 Body cells sometimes produce hydrogen peroxide, H_2O_2.
Hydrogen peroxide is dangerous in the human body.
The enzyme, catalase, speeds up the breakdown of hydrogen peroxide.

hydrogen peroxide $\xrightarrow{\text{catalase}}$ water + oxygen

a) Why is catalase written above the reaction arrow? [1]

b) In what way is catalase different to manganese(IV) oxide, which can also break down hydrogen peroxide? [2]

3 Temperature affects particle movement.

a) Draw two diagrams to show reacting particles at:

i) A low temperature. [1]

ii) A high temperature. [1]

b) Explain why reactions are faster at a higher temperature. [3]

> Total Marks .... / 10

Equilibria

1 Which of the following reactions are **reversible**?

A $CH_4(g) + 2O_2(g) \rightarrow CO_2(g) + 2H_2O(l)$

B $2Li(s) + 2HCl(aq) \rightarrow 2LiCl(aq) + H_2(g)$

C $H_2(g) + I_2(g) \rightleftharpoons 2HI(g)$

D $NH_4Cl(s) \rightleftharpoons NH_3(g) + HCl(g)$ [2]

2 A dynamic equilibrium can be reached when there is a closed system.

Explain what is meant by the term **closed system**. [2]

3 Which of the following graphs shows a dynamic equilibrium?

A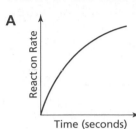

B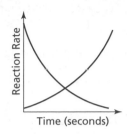

C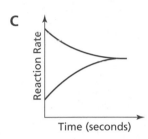

D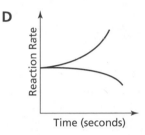

[1]

4 HT Give the **three** factors that can affect the position of equilibrium. [3]

5 HT Define Le Chatelier's principle. [2]

6 HT A reversible reaction that reaches dynamic equilibrium is shown below:

$$CH_3COOH(aq) \rightleftharpoons CH_3COO^-(aq) + H^+(aq)$$

This is an endothermic reaction in the forward direction (reactant → product).

a) What would happen to the amount of product if the temperature was
increased? [1]

b) What would happen to the amount of product if the pressure were increased? [1]

c) What would happen to the reaction if the amount of $CH_3COO^-(aq)$
was increased? [1]

7 HT Mark has set up an equilibrium reaction.
The equation for the reaction is:

$$H_2(g) + I_2(g) \rightleftharpoons 2HI(g)$$

Mark wants to maximise the amount of HI produced.
The reaction is exothermic in the direction reactants to product.

What variables can Mark alter to maximise HI production? [2]

Total Marks _____ / 15

Improving Processes and Products

You must be able to:

- Describe how to extract metals from their ores using carbon
- Explain the use of electrolysis to extract metals higher than carbon in the reactivity series
- **HT** Evaluate different types of biological metal extraction.

Industrial Metal Extraction Using the Reactivity Series

- It is only the least reactive metals that are found in their pure, elemental form (e.g. gold, platinum).
- All other metals react with other elements to form **minerals**, which are substances made up of metal compounds (typically oxides, sulfides or carbonates).
- An **ore** is a rock that contains minerals from which a metal can be extracted profitably.
- Carbon, when heated with a metal oxide, will **displace** a metal that is below it in the reactivity series.
- The products formed will be molten metal and carbon dioxide.
- An example is the extraction of copper from its ore.
- There are a number of different ores of copper. Each is a different copper compound.
- For example, cuprite is copper(I) oxide (Cu_2O) and chalcocite is copper(I) sulfide (Cu_2S).
- In the school laboratory, the displacement reaction can easily be demonstrated with pure copper(II) oxide.

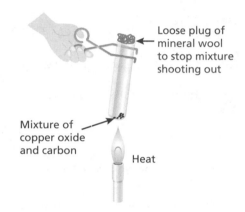

Loose plug of mineral wool to stop mixture shooting out

Mixture of copper oxide and carbon

Heat

> copper(II) oxide + carbon ⟶ copper + carbon dioxide

- The copper(II) oxide used in laboratories is free of impurities, which means displacement can take place more effectively.
- In industry, the reactions are scaled up to maximise the amount of metal that can be extracted.

Key Point

Carbon is a non-metal. However, in displacement reactions, it can displace less reactive metals as if it were one.

Industrial Metal Extraction Using Electrolysis

- Metals that are higher than carbon in the reactivity series cannot be displaced from their compounds by carbon.
- Industrial chemists use **electrolysis** to extract metals instead.
- The ore is heated to high temperatures to make it molten so that the ions move freely.
- The mixture is poured into an electrolytic cell with electrodes made of carbon.
- A typical example is the extraction of aluminium from its ore:
  - The aluminium ore is bauxite, which consists mainly of aluminium oxide (Al_2O_3).
  - The reaction is: aluminium oxide → aluminium + oxygen

Aluminium Electrolysis

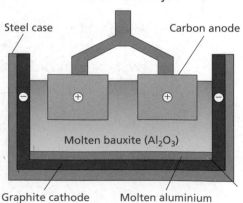

Steel case

Carbon anode

Molten bauxite (Al_2O_3)

Graphite cathode

Molten aluminium

| HT At the cathode: | $Al^{3+}(l) + 3e^- \longrightarrow Al(l)$ |
| At the anode: | $2O^{2-}(l) \longrightarrow O_2(g) + 4e^-$ |

- The oxygen reacts with the carbon anode to form carbon dioxide.
- This wears the anode out, so it has to be replaced regularly.

HT Biological Extraction Methods

- In recent years, scientists have developed ways of using living organisms to extract metals.
- **Bacteria**:
  - Bacteria are single-celled organisms that reproduce rapidly.
  - They can be bred to survive in very high concentrations of metal ions.
  - As they feed, they accumulate the ions in their cells.
  - Chemists can then extract the ions from the bacteria.
  - As the bacteria accumulate a specific ion, they will ignore the waste material.
  - This means that this process is very efficient.
  - It is much cheaper than extracting metals in a furnace (by displacement).
  - However, once the ions are removed from the ore there is a risk of them leaking from the bacteria into the environment.
- **Phytoextraction**:
  - Heavy metals, such as lead, are dangerous to most living organisms in high concentrations.
  - Plants cannot normally tolerate high levels of metal ions.
  - However, plants can be bred, or genetically engineered, to accumulate heavy metal ions. These are called **hyperaccumulators**.
  - The plants take in the ions via the roots and store them in non-essential tissues.
  - The willow tree has been used in this way to remove the heavy metal cadmium from contaminated areas.
  - Advantages are that the process is environmentally friendly – the soil is not harmed in the process – and it is cheap.
  - The disadvantage is that it takes a while for the plants to grow, so it is a long-term project.

 Key Point

As scientists learn how organisms grow and respond to the environment, they can start creating new technologies to address problems such as pollution.

Quick Test

1. Galena is an ore containing lead. Suggest an appropriate method for extracting the lead.
2. Why do the anodes have to be replaced regularly in the electrolysis of molten aluminium?
3. HT Give **two** reasons why bacteria are used for metal extraction.

 Key Words

minerals
ore
displace
electrolysis
HT **phytoextraction**
HT **hyperaccumulators**

The Haber Process

You must be able to:

- Explain the importance of the Haber process
- Explain why there is a trade-off between rate of production and position of equilibrium in some industrial processes.

The Haber Process

- Ammonia is used to manufacture a huge range of different chemicals, from dyes and explosives to drugs and nitric acid.
- Its biggest use is in the production of cheap, nitrogen-based fertilisers.
- Nitrogen is very unreactive.
- Although the atmosphere is 78% nitrogen, only a few plants contain bacteria that can convert the nitrogen into ammonium compounds.
- Plants need nitrogen to make protein and grow, so scientists had to find a way of converting the nitrogen into a more useable form.
- Converting nitrogen into ammonia and then into other compounds was the solution.
- Ammonia is made on a very large scale using the **Haber process**.
- The reactants are nitrogen (from the air) and hydrogen (from natural gas or the cracking of crude oil). The reaction is **reversible**:

> **nitrogen + hydrogen ⇌ ammonia**

> ### Key Point
>
> The Haber process is used to produce ammonia on a large scale. Ammonia is used to make fertilisers, which help plants to grow faster and bigger, increasing the amount of food produced.

HT Trade-Offs

- The chemical equation for the Haber process is:

$$N_2(g) + 3H_2(g) \rightleftharpoons 2NH_3(g)$$

- This reaction reaches a **dynamic equilibrium**.
- Factors affecting the equilibrium are temperature, concentration and pressure.
- For commercial reasons, it is important that the maximum amount of ammonia is made in the shortest possible time and at a reasonable cost.
- There will always be compromises involved in the industrial manufacture of chemicals.
- The Haber process reaction is exothermic—a low temperature increases the yield, but then the reaction is too slow.
- There are four molecules on the left-hand side of the equation, but only two on the right.
- Therefore, a high pressure increases the yield. However, it is very expensive to build factories and equipment that can withstand high pressures.

- A **catalyst** increases the rate of reaction but does not alter the position of the equilibrium.
- A compromise is reached with the following conditions:
  - temperature of 450°C
  - pressure of 200 atmospheres
  - catalyst of iron.
- This gives a reaction with an acceptable percentage yield.
- The **contact process** is used to manufacture sulfuric acid, another key substance that can be used to make a variety of other chemicals.
- The chemical equation for the contact process is:

$$2SO_2(g) + O_2(g) \rightleftharpoons 2SO_3(g)$$

- It is an exothermic reaction.

Interpreting Graphs

- The graph (right) shows how temperature and pressure affect the rate of reaction in the Haber process.
- The information given allows you to work out that:
  - the yield falls when temperature is increased
  - the yield increases as pressure increases.

Factors Affecting Manufacture

- The cost of making a new substance depends on:
  - the price of energy (gas and electricity)
  - labour costs (wages for employees)
  - how quickly the new substance can be made (cost of catalyst)
  - the cost of starting materials (reactants)
  - the cost of equipment needed (plant and machinery).
- Factors that affect the cost of making a new substance include:
  - the pressure – the higher the pressure, the higher the plant cost
  - the temperature – the higher the temperature, the higher the energy cost
  - the catalysts – catalysts can be expensive to buy, but production costs are reduced because they increase the rate of reaction
  - the number of people needed to operate machinery – **automation** reduces the wages bill
  - the amount of unreacted material that can be recycled – recycling reduces costs.

Revise

Key Point

You need to be able to interpret graphs of reaction conditions versus rate of reaction.

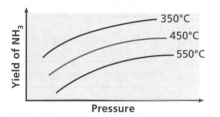

Key Point

Chemical costs vary according to the purity. Even so, when bought in bulk the overall cost can be reduced.

Key Words

Haber process
reversible
HT **dynamic equilibrium**
HT **catalyst**
HT **contact process**
HT **automation**

Quick Test

1. Name the **two** reactants in the Haber process.
2. HT Why isn't the Haber process carried out at the highest possible pressure?
3. HT Give **three** factors that affect the cost of making a new substance.

Life Cycle Assessments, Recycling and Alloys

You must be able to:

- Describe the differences between industrial and laboratory scale production
- Describe a life cycle assessment for a material or a product
- Explain recycling and evaluate when it is worth undertaking
- Describe the make-up and properties of key alloys.

Scales of Production

- Manufacturing chemicals on an industrial scale is very different to producing chemicals in a laboratory.

Laboratory Production	Industrial Production
Small quantities of chemicals used	Large quantities of chemicals used
Chemicals are expensive to buy	Bulk buying means chemical are cheaper
Energy usage is not an issue	There is a need to reduce costs and, therefore, energy usage
Equipment is usually glassware	Large-scale expensive factory equipment is needed
Chemists work on each reaction	Automation is common

Life Cycle Assessments

- A **life cycle assessment** is used to determine the environmental impact of a product throughout its life cycle.
- A product can be viewed as having four stages in its life cycle:
  1. Obtaining raw materials
  2. Manufacture
  3. Use
  4. Disposal.
- For each stage, the assessment looks at how much energy is needed, how **sustainable** the processes are, and what the environmental impact is.

Plastic Bags	Paper Bags
Plastic bags are made out of plastic.	Paper bags are made from wood pulp.
The plastic is made from crude oil, which is not **renewable**.	The wood pulp is made from trees, which are renewable.
Making the plastic requires a lot of energy (for fractional distillation and cracking) and is not sustainable.	This means that paper bags are sustainable (low environmental impact).
The environmental impact of making the bags is high, as oil is running out.	Paper bags are often only used once. Therefore, they are disposed of in large numbers.
Plastic bags are often only used once, are non-biodegradable and disposed of in landfill sites.	Paper bags can be recycled and repulped to make new bags.

Recycling

- Recycling is the process of taking materials and using them to make new products, e.g. making a vase from a glass bottle.
- Recycling materials means:
    - less quarrying for raw materials is required
    - less energy is used to extract metals from ores
    - the limited ore and crude oil reserves will last longer (saves natural resources)
    - disposal problems are reduced.
- It is important to evaluate whether an object should be recycled or not.
- Sometimes recycling could be more environmentally damaging than just disposal.
- For example, plastic objects that are placed in recycling bins may have contaminants – they would have to be deep cleaned, which the recycling companies can't do due to high costs.

Alloys

- **Alloys** are a mixture of elements, containing at least one metal.
- An alloy has different properties to the individual elements.
- Steel is an alloy – it is a mixture of iron, carbon and other metals (depending on the type of steel).
- Stainless steel is resistant to corrosion and is used to make cutlery.
- The elements in some alloys stop the metal atoms from slipping past one another when a force is applied (making it stronger).
- For example, in steel the carbon stays in the gaps between the iron atoms – it is an **interstitial** alloy.
- Brass is a mixture of copper and zinc.
- Mixing these elements in different proportions produces brasses with different properties.
- Brass can easily be bent and made into different shapes.
- It is used to make musical instruments, jewellery, zips and coins.
- Zinc and copper atoms are a similar size.
- In brass, the zinc and copper atoms are randomly mixed together – it is a **substitutional** alloy.
- Bronze is an alloy of copper and tin. It is harder and more durable than copper alone.
- Bronze is used for making weapons and statues.
- Solder is an alloy of lead and tin. It melts at a low temperature.
- Solder is used to join electrical parts to circuit boards.
- Duralumin is an alloy of aluminium, copper, magnesium and small amounts of manganese.
- It is a very light, strong alloy, used in the production of aircraft.

> **Key Point**
>
> Being environmentally friendly is seen as an advantage when marketing and selling a product.

> **Key Point**
>
> The atoms of the different elements in an alloy are not chemically bonded together – an alloy is a mixture.

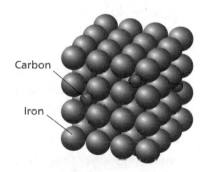

Carbon

Iron

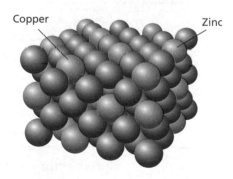

Copper · Zinc

> **Quick Test**
>
> 1. Why is returning glass bottles to a store for refilling *not* an example of recycling?
> 2. Suggest why some plastics are not suitable for recycling.
> 3. What are the main differences between the structures of the alloys steel and brass?

> **Key Words**
>
> life cycle assessment
> **sustainable**
> **renewable**
> **alloy**
> **interstitial**
> **substitutional**

Using Materials

You must be able to:

- Compare the properties of a range of materials
- Explain how metals corrode
- Suggest ways to prevent or reduce the corrosion of metals.

Properties of Materials

- Materials are chosen for a specific purpose because of their physical properties, e.g. teapots are made from materials that can hold boiling water without melting.

Metals

- Metals are generally easy to manipulate and extract and can be made into alloys with different properties to the original metals.
- Some metals are very hard and strong.
- For example, iron (when mixed with other elements to form steels) is used in cars, machine tools and buildings.
- Others metals, such as copper and silver, are **ductile** and can be used to make wires.

Corrosion

- Most metals will react with oxygen to form a metal oxide. This process is called **corrosion**.
- Corrosion is an expensive problem, because corroded metal has to be replaced.
- The corrosion of iron (an oxidation reaction) is called **rusting** – it costs billions of pounds per year worldwide.

> **iron + oxygen + water ⟶ hydrated iron(III) oxide**

- Rusting occurs even faster when iron comes into contact with salt water or acid rain.
- Aluminium is very reactive, but it does not corrode in air and water.
- Atoms on the surface quickly react to form aluminium oxide, which creates a protective barrier.

> **aluminium + oxygen ⟶ aluminium oxide**

- There are two ways of preventing corrosion:
  - a physical barrier between the metal and water and oxygen
  - **sacrificial protection**.
- Choosing a material to protect a metal depends upon the intended use.
- Painting metal is a common solution (e.g. an iron bridge), as long as the paint is not likely to get chipped.
- Coating metal in plastic is more durable (e.g. chain linked fences).
- Oil and grease is used on objects that move (e.g. bicycle chains).

Key Point

Corrosion damages metals and can make objects unsafe. Finding cheap and effective ways to reduce corrosion preserves the metal and saves money.

- Sacrificial protection involves attaching a more reactive metal to the surface of the metal being protected, e.g. attaching zinc or magnesium to iron.
- With the hulls of ships, a block of zinc is attached.
- The more reactive metal reacts and corrodes first, protecting the main metal.
- Galvanising is the process of covering a metal with a layer of zinc.
- This provides a physical barrier and sacrificial protection.

Glass and Clay Ceramics

- Glass is a non-crystalline solid that is transparent or translucent.
- It is made by heating silica with lime and sodium carbonate.
- Glass is unreactive, so it can be used to make bottles to store reactive chemicals, windows and tableware.
- Clay **ceramics** are made by shaping clay and then heating at a high temperature in an oven.
- Fine clays are used to make plates in a dinner service.
- Clays with a larger particle size are used to make the separators that provide electrical insulation in power lines.

Key Point

There is an extremely large variety of materials available. However, not all will be environmentally friendly and meet the requirements of a life cycle assessment.

Polymers

- There is a huge range of polymers that are created for different purposes.

Polymer	Properties	Uses
Polythene or poly(ethene)	• Light • Flexible • Easily moulded • Can be printed on	• Plastic bags – the plastic is flexible and light. • Moulded containers – the plastic is easily moulded.
Polystyrene	• Light • Poor conductor of heat	• Insulation – the plastic is a poor conductor of heat.
Polyester	• Lightweight • Waterproof • Tough • Can be coloured	• Clothing – the plastic can be made into fibres, is lightweight, tough, waterproof and can be coloured. • Bottles – the plastic is lightweight and waterproof.

Composites

- **Composites** are mixtures or layers of different materials that are chemically bonded together.
- They are often very strong and durable.
- Examples of composites include:
  - concrete (a mixture of aggregates, sand and cement)
  - carbon composites, used for re-entry shields on spacecraft
  - Kevlar™, used to make bulletproof vests.

Key Words

ductile
corrosion
rusting
sacrificial protection
ceramics
composites

Quick Test

1. Why are blocks of zinc attached to ships' hulls?
2. Why are iron nails galvanised?
3. Suggest why polystyrene is used to make cups.

Organic Chemistry

You must be able to:

- Identify and draw the first four members of the alkanes and the alkenes
- Identify and draw the first four members of the alcohols
- Identify and draw the first four members of the carboxylic acids.

Organic Compounds

- Carbon can form up to four separate covalent bonds with other atoms.
- Compounds containing carbon are called **organic** compounds.
- There are a few exceptions, e.g. oxides of carbon and carbonates.
- Organic compounds are divided into groups of compounds with similar properties.
- Each group is called a **homologous series**.
- Compounds in the same homologous series:
  - can be represented by a general formula
  - have similar chemical properties, because they contain the same **functional group**
  - can be produced by similar reactions
  - show a regular change (a pattern) in their physical properties.
- There are four main groups that you will be tested on: **alkanes**, **alkenes**, **alcohols** and **carboxylic acids**.

Key Point

Always check the number of covalent bonds on a carbon atom. There should be four.

Key Point

The carbon–hydrogen bond stores energy. The longer the chain of molecules, the greater the amount of stored energy.

Alkanes

- **Hydrocarbons** are molecules that are made up of only hydrogen and carbon.
- Alkanes are a homologous series of hydrocarbons.
- They are described as 'saturated hydrocarbons' because they contain only single carbon–carbon covalent bonds.
- The name of an alkane always ends in -*ane*.
- The general formula for alkanes is C_nH_{2n+2} (the *n* stands for the number of carbon atoms).
- Each member of the series differs from the next by one $-CH_2-$ unit.
- The table shows the displayed and **molecular formulae** for the first four members of the alkane series:

Alkane	Methane, CH_4	Ethane, C_2H_6	Propane, C_3H_8	Butane, C_4H_{10}
Displayed Formula	H | H—C—H | H	H H | | H—C—C—H | | H H	H H H | | | H—C—C—C—H | | | H H H	H H H H | | | | H—C—C—C—C—H | | | | H H H H

Alkenes

- Alkenes are a homologous series of hydrocarbons.
- Alkenes are described as being unsaturated because they contain one or more double carbon=carbon covalent bond, i.e. the carbon atom is not bonded to the maximum number of atoms.

- The name of an alkene always ends in *-ene*.
- The general formula for alkenes is C_nH_{2n}.

Alkene	Ethene, C_2H_4	Propene, C_3H_6	Butene, C_4H_8	Pentene, C_5H_{10}
Displayed Formula	(displayed structure)	(displayed structure)	(displayed structure)	(displayed structure)

Alcohols and Carboxylic Acids

- Alcohols are a homologous series of organic compounds that have the hydroxyl (–OH) functional group.
- The general formula for an alcohol is $C_nH_{2n+1}OH$.
- With long-chained alcohols, it is common practice to write the **structural formula**.

Alcohol	Methanol	Ethanol	Propanol	Butanol
Displayed Formula	(displayed structure)	(displayed structure)	(displayed structure)	(displayed structure)
Molecular Formula	CH_4O	C_2H_6O	C_3H_8O	$C_4H_{10}O$
Structural Formula	CH_3OH	CH_3CH_2OH	$CH_3CH_2CH_2OH$	$CH_3CH_2CH_2CH_2OH$

- Carboxylic acids are a homologous series of organic acids which have the carboxyl (–COOH) functional group.
- The general formula for a carboxylic acid is $C_nH_{2n+1}COOH$.

Carboxylic Acid	Methanoic acid	Ethanoic acid	Propanoic acid	Butanoic acid
Displayed Formula	(displayed structure)	(displayed structure)	(displayed structure)	(displayed structure)
Molecular Formula	CH_2O_2	$C_2H_4O_2$	$C_3H_6O_2$	$C_4H_8O_2$
Structural Formula	$HCOOH$	CH_3COOH	CH_3CH_2COOH	$CH_3CH_2CH_2COOH$

Key Words

organic
homologous series
functional group
alkanes
alkenes
alcohol
carboxylic acid
hydrocarbon
molecular formula
structural formula

Quick Test

1. What is the molecular formula for propene?
2. What is the general formula for the alcohol homologous series?
3. What functional group is present in carboxylic acids?

Organic Compounds and Polymers

You must be able to:

- Describe the reactions of alkanes, alkenes and alcohols
- Recall the basic principles of addition polymerisation
- **HT** Explain the principles of condensation polymerisation.

Combustion

- When alkanes, alkenes and alcohols are burned in excess oxygen, they produce carbon dioxide and water vapour.

> alkane / alkene / alcohol + oxygen $\longrightarrow$ carbon dioxide + water vapour

- As the number of carbon atoms in the compounds increases, carbon monoxide and carbon are more likely to be produced.

Bromination and Hydrogenation

- A simple test to distinguish between alkenes and alkanes is to add bromine, a brown coloured liquid.
- Alkenes decolourise bromine, forming a colourless liquid.
- This reaction is known as bromination.
- Alkanes have no effect on bromine.
- When an alkene reacts with hydrogen:
  - the double carbon=carbon bonds are broken
  - the carbon atoms bond with the maximum number of hydrogen atoms possible.
- This saturates the compound, converting the alkene to an alkane.
- The reaction is known as hydrogenation. For example:

> ethene + hydrogen $\longrightarrow$ ethane
>
> $C_2H_4(g) + H_2(g) \longrightarrow C_2H_6(g)$

- Hydrogenation and bromination are addition reactions – an unsaturated organic compound combines with another substance to form a single compound.

Oxidising Alcohol

- Potassium manganate(VII) is a very powerful, dark purple **oxidising agent**.
- When added to alcohol and heated, a carboxylic acid forms and the purple colour disappears.

> alcohol + potassium manganate(VII) $\longrightarrow$ carboxylic acid + manganese(IV) oxide + potassium hydroxide + water

Key Point

Alkanes, alkenes and alcohols react differently in chemical reactions, except for combustion.

Key Point

Organic molecules will react with other reactants. The products that are formed vary depending on the length of the chain.

Polymers

- **Polymers** are long-chained molecules made up of repeating subunits, called **monomers**.
- Monomers with different functional groups produce polymers with different properties.
- Alkene monomers join together to make a polymer in a **polymerisation** reaction.
- This process needs high pressure and a catalyst.
- Displayed formulae can be used to show a polymerisation reaction.
- However, the standard way of displaying a polymer formula is:

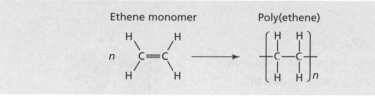

- If you are given a polymer formula, you should be able to show what the monomer unit is:

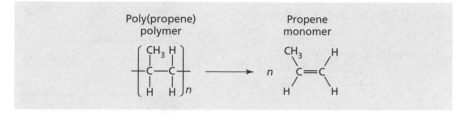

- When monomer units are joined together like this, it is called **addition polymerisation**.
- Polymers, such as proteins, DNA, starch and cellulose, are found in nature.
- The nucleic acid, DNA, is a polymer made up of repeating nucleotides.
- A nucleotide is a phosphate, a pentose sugar and a base.
- The nucleotides in DNA are called cytosine, guanine, adenine and thymine.
- Some natural polymers use repeating sugar monomers (e.g. starch or glycogen).
- Proteins are polymers made out of amino acid monomers.

HT Condensation Polymerisation

- **Condensation polymerisation** is when monomer units join together and a small molecule, such as water, is removed.
- Nylon is made through a condensation reaction.
- There are two monomers that react together to make nylon: a dicarboxylic acid and a diamine.
- A dicarboxylic acid has a COOH group at each end of the chain.
- An **amine** has a NH$_2$ group at each end.

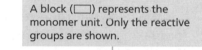

A block (▭) represents the monomer unit. Only the reactive groups are shown.

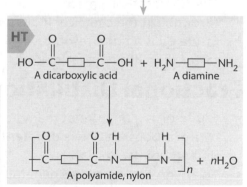

Quick Test

1. Potassium manganate(VII) is added to a colourless solution. The purple colour disappears. Suggest what the solution could be.
2. HT What is a condensation polymerisation reaction?
3. Explain why proteins are polymers.

Key Words

oxidising agent
polymers
monomer
polymerisation
addition polymerisation
HT condensation
    polymerisation
HT amine

Crude Oil and Fuel Cells

You must be able to:

HT Describe how to make a polymer by addition and condensation

- Explain the processes of fractional distillation and cracking
- Describe alternatives to crude oil as a fuel source.

Producing Nylon

HT Making Polymers

- To make an addition polymer, such as poly(ethene):
  - The monomer units need to react in the presence of a **catalyst**.
  - The catalyst starts the reaction and causes a **chain reaction**.
- To make a condensation polymer, such as nylon:
  - Decanedioyl dichloride in cyclohexane is floated on a solution of diaminohexane in water.
  - Polymerisation occurs at the interface (the boundary between the two solutions).
  - Pulling out the polymer that forms results in a long thread of nylon.

Crude Oil

- Crude oil is a fossil fuel that formed over millions of years in the Earth's crust.
- It has become crucially important to our modern way of life.
- It is the main source of chemicals for the petrochemical industry, for making plastics and fuels such as petrol and diesel.
- The hydrocarbons present in crude oil are used extensively throughout the chemical industry.
- Crude oil is a finite resource, which means it is being used up much faster than it is being replaced.

> **Key Point**
>
> It is important to recognise that the components of crude oil can be changed into many other chemicals, including fuel, medicines and plastics.

Fractional Distillation

- Crude oil is a mixture of many hydrocarbons that have different boiling points.
- The hydrocarbon chains generally all belong to the alkane homologous series (general formula C_nH_{2n+2}).
- The longer the chain, the greater the intermolecular forces, so the higher the boiling point.
- This means that crude oil can be separated into useful fractions (parts) that contain mixtures of hydrocarbons with similar boiling points.
- The process used is **fractional distillation**.

Fractional Distillation Tower

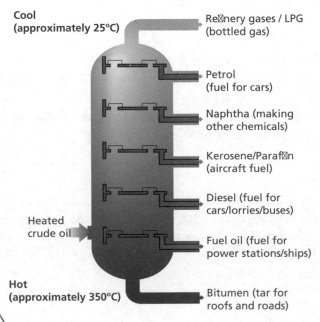

Cool (approximately 25°C)

Refinery gases / LPG (bottled gas)

Petrol (fuel for cars)

Naphtha (making other chemicals)

Kerosene/Paraffin (aircraft fuel)

Diesel (fuel for cars/lorries/buses)

Heated crude oil

Fuel oil (fuel for power stations/ships)

Hot (approximately 350°C)

Bitumen (tar for roofs and roads)

Small molecules

Low boiling point

Evaporate easily

Burn easily

Large molecules

High boiling point

Don't evaporate easily

Don't burn easily

- The crude oil is heated in a fractionating column.
- The column has a temperature gradient – it is hotter at the bottom of the column than at the top:
    - Fractions with low boiling points leave towards the top of the fractionating column.
    - Fractions with high boiling points leave towards the bottom of the fractionating column.

Cracking

- There is not enough petrol in crude oil to meet demand.
- A process called **cracking** is used to change the parts of crude oil that cannot be used into smaller, more useful molecules.
- Cracking requires a catalyst, high temperature and high pressure.
- Petrol can be produced in this way, as can other short-chain hydrocarbons used to make polymers for plastics and fuels.
- For example, C_6H_{14} can be cracked to form C_2H_4 (ethene) and C_4H_{10} (butane).

Battery and Fuel Cells

- As reserves of crude oil are being used up, scientists are investigating other sources of energy.
- Battery cells use the electrolysis of two different metals and an electrolyte:
    - Electrons move from the more reactive metal (in this case, zinc), to the less reactive metal (copper).
    - Over time, the more reactive metal (zinc) is used up, the reaction stops and the battery becomes dead.
    - Batteries are used as a power source in portable electrical devices.
- A **fuel cell** is a device that converts chemical energy (stored in covalent bonds) directly into electrical energy.
- The fuel cell produces a potential difference (a voltage) that can be used in electrical circuits.
- It is much more efficient than a battery cell.
- One example of a fuel cell is the hydrogen / oxygen cell.
- A disadvantage is that the fuel has to be continually added, so they are not suitable for portable electrical devices.

> **hydrogen + oxygen $\longrightarrow$ water (+ electrical energy)**

Advantages	Disadvantages
No harmful **emissions**	Difficult to store hydrogen safely
Hydrogen is easy to produce	Expensive
Hydrogen has a high energy efficiency	Difficult to move it around

A Simplified Battery Cell

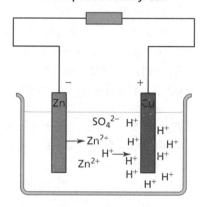

A Fuel Cell

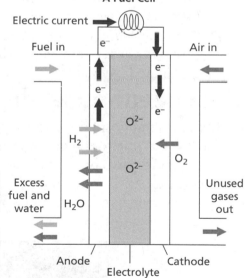

Quick Test

1. What is meant by the term 'cracking'?
2. Why do longer-chain hydrocarbons have higher boiling points?
3. Give **three** disadvantages of hydrogen fuel cells.

Interpreting and Interacting with Earth's Systems

You must be able to:

- Describe how the Earth's atmosphere has developed over time
- Explain the greenhouse effect
- Describe the effects of greenhouse gases.

The Earth's Atmosphere

- The current composition of the Earth's **atmosphere** is 78% nitrogen, 21% oxygen and 1% other gases, including 0.040% (400ppm) carbon dioxide.
- The Earth's atmosphere has not always been the same as it is today – it has gradually changed over billions of years.
- A current theory for how the Earth's atmosphere evolved is:
  1. A hot, volcanic Earth released gases from the crust.
  2. The initial atmosphere contained ammonia, carbon dioxide and water vapour.
  3. As the Earth cooled, its surface temperature gradually fell below 100°C and the water vapour condensed into liquid water.
  4. Some carbon dioxide dissolved in the newly formed oceans, removing it from the atmosphere.
  5. The levels of nitrogen in the atmosphere increased as nitrifying bacteria released nitrogen.
  6. The development of primitive plants that could photosynthesise removed carbon dioxide from the atmosphere and added oxygen.

The Greenhouse Effect

- The Earth's atmosphere does a very good job at keeping the Earth warm due to the **greenhouse effect**.
- Greenhouse gases prevent **infrared radiation** from escaping into space.
- Carbon dioxide, methane and water vapour are powerful greenhouse gases.

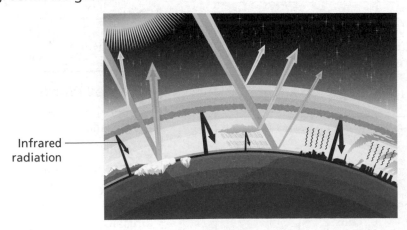

Infrared radiation

> ### Key Point
>
> The atmosphere is very thin. If the Earth were the size of a basketball, the atmosphere would only be as thick as a layer of clingfilm wrapped around it.

> ### Key Point
>
> Scientists take a variety of samples from all over the planet and analyse them to build models of the gases that were present in the atmosphere at different times in Earth's history.

> ### Key Point
>
> Scientific consensus (over 97% of scientists globally) is that it is extremely likely humans are the main cause of climate change.

Climate Change and the Enhanced Greenhouse Effect

- Evidence strongly indicates that humans have contributed to increasing levels of carbon dioxide, creating an **enhanced greenhouse effect**.
- Three key factors have affected the balance of carbon dioxide in the atmosphere:
  1. Burning **fossil fuels**, which releases CO_2 into the atmosphere.
  2. Deforestation over large areas of the Earth's surface, which reduces photosynthesis, so less CO_2 is removed from the atmosphere.
  3. A growing global population, which directly and indirectly contributes to the above factors.
- Carbon dioxide levels have increased from approximately 0.028%, at the time when humans first started using industrial technology, to the 0.040% seen today.
- As a result, the average global temperatures are changing, bringing about changes in weather patterns (climate change) over the entire planet.
- The evidence points to humans being the main cause of global climate change. However, these are still controversial issues.
- There are natural sources of greenhouse gases, e.g. volcanic eruptions, carbon dioxide emissions from the oceans and respiration.

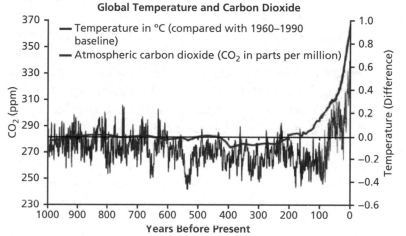

Global Temperature and Carbon Dioxide

— Temperature in °C (compared with 1960–1990 baseline)
— Atmospheric carbon dioxide (CO_2 in parts per million)

Years Before Present

- Detailed global temperature and CO_2 measurements were not made in the past, so scientists have to take samples and model what the atmosphere was like at different times.
- There is still a lot of research being carried out to gather more evidence and improve the climate models.
- To slow or stop climate change, we need to:
  - Reduce our reliance on fossil fuels, and find and use alternative energy sources.
  - Be more aware of our energy use and not waste energy.
  - Implement **carbon capture** schemes, such as planting forests – the forests store carbon whilst they are growing.
  - Reduce waste sent to landfill – waste often produces methane gas, which has a greater greenhouse effect than carbon dioxide.

Quick Test

1. What was the main source of carbon dioxide, methane and ammonia in the Earth's early atmosphere?
2. What process led to the large amount of oxygen in the atmosphere?
3. Which greenhouse gas is produced by landfill sites?

Key Words

atmosphere
infrared radiation
greenhouse effect
enhanced greenhouse effect
fossil fuels
carbon capture

Air Pollution, Potable Water and Fertilisers

You must be able to:

- Describe the effects of CO, SO_2, NO_x and particulates in the atmosphere
- Describe methods for increasing potable water
- Describe how NPK fertilisers are made.

Carbon Monoxide, Sulfur Dioxide and Oxides of Nitrogen (NO_x)

- When hydrocarbons are combusted with oxygen, they will produce carbon dioxide (CO_2) and water.
- If there is not enough oxygen, carbon monoxide (CO) is produced.
- Carbon monoxide is toxic.
- It bonds more tightly to haemoglobin than oxygen does, starving the blood of oxygen so that the body suffocates slowly.
- Gas appliances need to be checked regularly for this reason.
- Burning fossil fuels produces large amounts of carbon monoxide.
- Sulfur dioxide (SO_2), in small quantities, is used as a preservative in wine to prevent it from going off.
- Sulfur dioxide is produced when fossil fuels are burned.
- It is also produced in large quantities during volcanic eruptions.
- Sulfur dioxide dissolves in water vapour in the atmosphere causing acid rain, which kills plants and aquatic life, erodes stonework and corrodes ironwork.
- SO_2 also reacts with other chemicals in the air forming fine **particulates** (see below).
- Nitrogen and oxygen from the air react in a hot car engine to form nitrogen monoxide (NO) and nitrogen dioxide (NO_2).
- NO_x is used to refer to the various nitrogen oxides formed.
- They can cause photochemical smog and acid rain.

> ### Key Point
>
> CO attaches to the haemoglobin because the shape of the CO molecule is so similar to the O_2 molecule.

Particulates

- Particulates are very small, solid particles that can cause lung problems and **respiratory** diseases.
- If a fuel is burned very inefficiently, carbon (soot) is produced.
- As well as causing respiratory problems, soot can coat buildings and trees in a black layer.
- Realising the dangers of **air pollutants**, governments have passed laws dictating the maximum levels that are allowed to be emitted, with the aim of improving air quality.

Potable Water

- Although the surface of the planet is covered in water, much of it is not safe to drink.
- Approximately 26% of the world's population only has access to unsafe water (e.g. contaminated by faeces).
- Providing potable water (water that is safe to drink) is an important goal for many countries.

Treating Waste Water

- When rain water goes down a drain, the toilet is flushed or a bath is emptied, the water enters the sewage system.
- This water is treated at a sewage treatment works before being piped to the tap:
  1. The water is filtered to remove large objects.
  2. The water is held in a **sedimentation tank** to remove small particulates.
  3. Bacteria are added to break down the particulates.
  4. Fine particles are removed by filtration.
  5. Chlorine is added to kill bacteria.
  6. The water is **ultra-filtered** through special membranes and stored in reservoirs.
  7. It is then treated before being piped back to the tap.

Desalination

- The majority of the world's water is in the seas and oceans.
- The water contains dissolved salt, which means it is undrinkable.
- **Desalination** (removing the salt to make the water drinkable) is an expensive process.
- There are two ways this can be achieved:
- **Evaporation / condensation:**
  1. Heat the salt water to boiling point to evaporate the water off.
  2. Cool the water vapour to recondense it into liquid form.
- **Reverse osmosis:**
  1. The salt water is filtered and passed into an osmosis tank.
  2. The salt water is forced against a partially permeable membrane.
  3. The membrane only lets water pass through, separating out the salt molecules.
  4. The water has chemicals added (e.g. chlorine) to kill bacteria.
- Both processes are very expensive due to their high energy needs.

The Industrial Production of Fertilisers

- There is a shortage of land suitable for growing crops.
- Chemists can create fertilisers to enrich soils to make them suitable for crop growth.
- The three main elements found in fertilisers that replace nutrients in the soil are nitrogen (N), phosphorus (P) and potassium (K).
- A variety of chemical pathways are used to produce fertilisers.
- Ammonia sulfate is made by neutralising sulfuric acid from the contact process (see page 79) with ammonia from the Haber process (see page 78).

 Key Point

All water molecules are identical, even if they have passed through other organisms.

Desalination

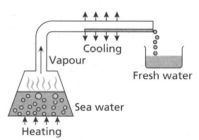

 Key Point

Fertilisers are made by neutralising an acid with ammonia or potassium hydroxide.

Key Words

particulates
respiratory
air pollutants
sedimentation tank
ultra-filtered
desalination
reverse osmosis

Quick Test

1. What is the biggest source of CO in the atmosphere?
2. Why is chlorine often added to drinking water?
3. What **two** processes are used to produce the reactants that make ammonium sulfate?

Review Questions

Predicting Chemical Reactions

1 Which of the following statements about a Group 1 element is **false**?

    A Conducts electricity

    B Is malleable

    C Dissolves in water to make bleach

    D Reacts with water to form hydrogen   [1]

2 Which of the following dot and cross diagrams could show a Group 0 element?   [1]

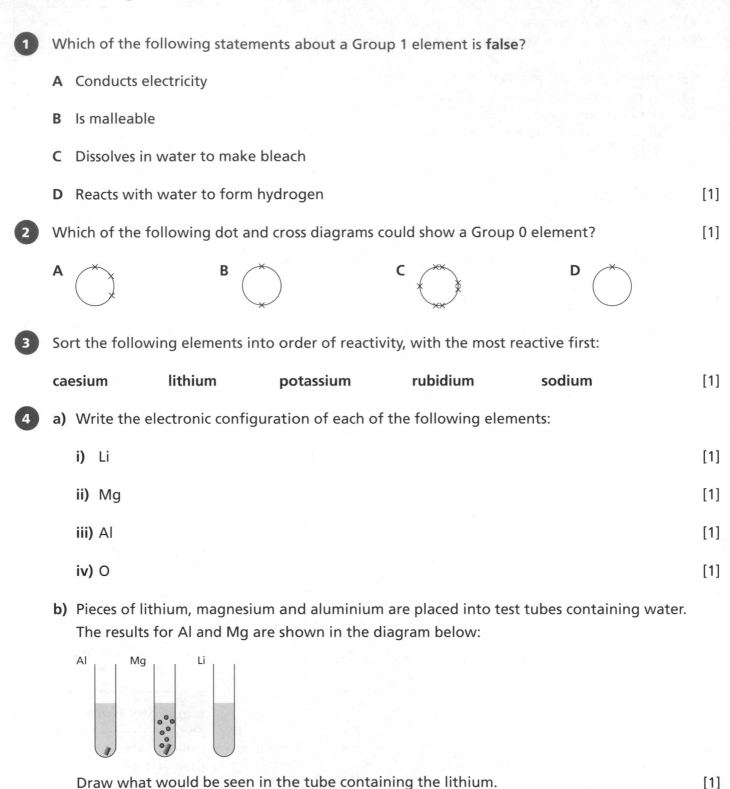

3 Sort the following elements into order of reactivity, with the most reactive first:

    caesium   **lithium**   **potassium**   **rubidium**   **sodium**   [1]

4 **a)** Write the electronic configuration of each of the following elements:

    i) Li   [1]

    ii) Mg   [1]

    iii) Al   [1]

    iv) O   [1]

  b) Pieces of lithium, magnesium and aluminium are placed into test tubes containing water. The results for Al and Mg are shown in the diagram below:

    Draw what would be seen in the tube containing the lithium.   [1]

  c) What gas is produced in the reaction?   [1]

Total Marks _____ / 9

Identifying the Products of Chemical Reactions

1 Ammonium chloride reacts with sodium hydroxide to produce ammonia, sodium chloride and water:

$$NH_4Cl(aq) + NaOH(aq) \rightarrow NH_3(aq) + NaCl(aq) + H_2O(l)$$

Write down the **ions** that take part in this reaction. [2]

2 Michael reacts a chemical, **X**, with sulfuric acid.
Michael then collects a gas in a test tube.
He places a lit splint over the end of the test tube and hears a squeaky pop.

Which of the following is the gas that has been produced?

A Hydrogen

B Carbon dioxide

C Oxygen

D Methane [1]

3 The decomposition of hydrogen peroxide in the presence of a catalyst produces oxygen.

Describe the test for oxygen gas. [2]

4 Arwen is testing a mystery solution to identify the ions that are present.
She thinks that the solution may contain iron(II) and sulfate ions.

Describe the tests that Arwen needs to carry out to prove these ions are present.
Include details of the results she would expect to see. [5]

5 HT Write ionic equations to show:

   a) Barium ions reacting with carbonate ions. [2]

   b) Calcium ions reacting with hydroxide ions. [2]

6 Describe how to test for the presence of chlorine gas. [2]

Total Marks _____ / 16

Review Questions

Ion Tests and Instrumental Methods of Analysis

1 Astra is carrying out experiments to identify gases.

Draw one line from each gas to the correct chemical test.

Name of Gas   **Chemical Test**

Turns moist red-litmus blue
Turns moist blue-litmus red
Turns acidified potassium manganate(VII) colourless
Turns limewater milky white
Burns with a squeaky pop
Relights a glowing splint

Carbon dioxide

Hydrogen

Oxygen

[3]

2 Claude is carrying out a series of flame tests.

a) Describe how to carry out a flame test. [4]

b) One of the samples that Claude examines burns with a red flame.

What ion may be present in the sample? [1]

3 **a)** Describe the test used to detect halide (halogen) ions in solution. [3]

b) A solution is thought to contain lithium bromide.
The test in part **a)** is carried out.

Write down the name and the colour of the precipitate that will form
if lithium bromide is present. [1]

c) HT Write the ionic equation for the reaction in part **b)**. [1]

4 Diane is measuring the pH of samples from a river.
She has an eye infection, which is making it difficult seeing the colour changes of the universal
indicator solution.

Suggest what she could do to get more accurate, quantitative results. [1]

Total Marks _____ / 14

Monitoring Chemical Reactions

1 Ethan carries out a titration of lithium hydroxide solution with dilute hydrochloric acid.
His results table is shown below:

Titration Number	1	2	3
Final Reading (cm³)	18.80	37.90	58.00
Initial reading (cm³)	0.00	20.00	40.00
Titre (cm³)			

a) Complete the table by adding the titre for Titrations **1**, **2** and **3**. [3]

b) Why should Ethan discard the measurement for Titration **1**? [1]

c) Calculate the mean titre for the experiment. [2]

d) HT Ethan used 30.0cm³ lithium hydroxide solution in the experiment.
The hydrochloric acid he used had a concentration of 0.200mol/dm³.

Calculate the concentration of the LiOH(aq) in mol/dm³. [4]

2 Look at the following titration curve:

a) On the graph, indicate with an **X** the
end point for the neutralisation. [1]

b) What does the end point represent? [1]

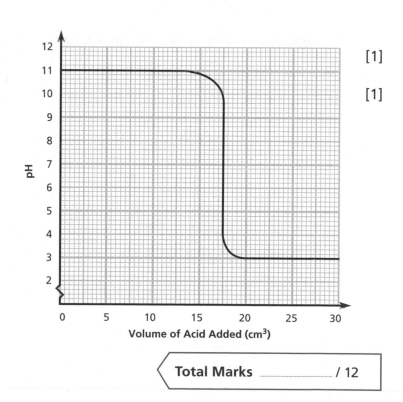

Volume of Acid Added (cm³)

Total Marks _____ / 12

Review Questions

Calculating Yields and Atom Economy

1 HT Kate carries out a titration of sodium hydroxide and nitric acid:

$$NaOH(aq) + HNO_3(aq) \rightarrow NaNO_3(aq) + H_2O(l)$$

She finds that $0.025dm^3$ of $0.2mol/dm^3$ HNO_3 is neutralised by $0.050dm^3$ of NaOH solution.

Calculate the concentration of NaOH solution. [3]

2 A balloon contains 3.00mol of helium gas at room temperature and pressure. The molar volume is $24dm^3$.

a) Calculate the volume of gas in the balloon. [1]

b) What would happen to the volume if the gas is cooled down? [1]

3 What is the correct formula for calculating the number of moles of gas?

A number of moles of gas = $\dfrac{\text{volume of gas}}{\text{molar volume}}$

B number of moles of gas = $\dfrac{\text{molar volume}}{\text{volume of gas}}$

C number of moles of gas = volume of gas × molar volume

D number of moles of gas = volume of gas [1]

4 A factory is manufacturing urea, $(NH_2)_2CO$, for use in the chemical industry. This involves reacting silver cyanate with ammonium chloride:

$$AgNCO(aq) + NH_4Cl(aq) \rightarrow (NH_2)_2CO(aq) + AgCl(aq)$$

a) 149.9kg of AgNCO(aq) is used.

How much urea should be produced? [1]

b) The percentage yield of the reaction was 80%.

Calculate the actual amount of $(NH_2)_2CO(aq)$ produced. [1]

Total Marks _____ / 8

Controlling Chemical Reactions

1 Jude is carrying out an investigation into how temperature affects the rate of reaction. He uses the reaction between sodium thiosulfate and hydrochloric acid:

$$Na_2S_2O_3(aq) + 2HCl(aq) \rightarrow 2NaCl(aq) + S(s) + SO_2(g) + H_2O(l)$$

The reactants in this reaction are colourless.
Sulfur produced in the reaction makes the solution turn an opaque yellow colour.
Jude uses a cross drawn under a beaker to help measure the reaction rate.

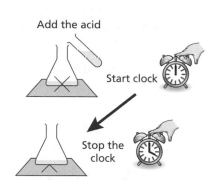

Jude carries out the reaction at three different temperatures using water baths.

a) Which of the following safety precautions should be taken during this experiment?

   A No naked flames

   B Keep the lab well ventilated

   C Wear eye protection

   D Keep a fire extinguisher nearby   [2]

b) The graph shows a reaction at two different temperatures.

Draw the line to show the results you would expect for 35°C.   [1]

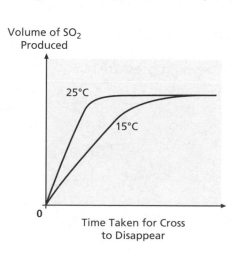

Total Marks _____ / 3

Review Questions

Catalysts and Activation Energy

1 Catalysts are often used in chemical reactions.
Modern car engines use a catalyst in the exhaust system.

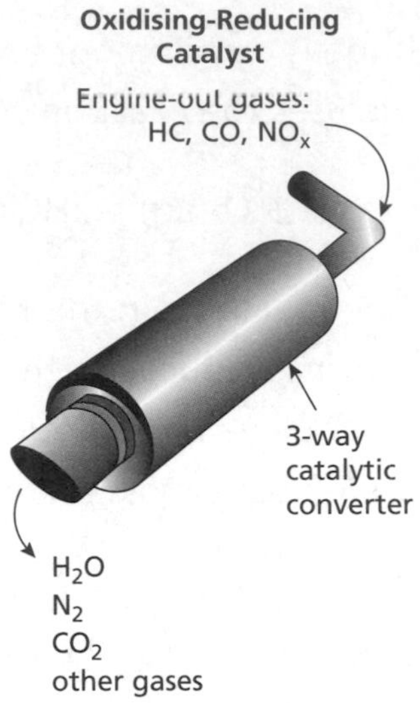

Oxidising-Reducing Catalyst

Engine-out gases: HC, CO, NO$_x$

3-way catalytic converter

H$_2$O
N$_2$
CO$_2$
other gases

a) What is a catalyst? [1]

b) Suggest why a car engine needs to have a catalytic converter. [2]

c) Draw a graph to show how the **activation energy** for a reaction is affected by the presence of a catalyst. [2]

2 Photosynthesis is represented by the equation:

$$6CO_2(g) + 6H_2O(l) \xrightarrow{\text{chlorophyll}} C_6H_{12}O_6(s) + 6O_2(g)$$

Why is chlorophyll included in this reaction?

A It is the pigment that makes leaves green.

B It is the site of photosynthesis.

C It interacts with catalysts (enzymes) in photosynthesis.

D It makes oxygen. [1]

3 The reaction between hydrogen and iodine can achieve a dynamic equilibrium when certain conditions are met.

Which of the following conditions have to be present for a **dynamic equilibrium** to occur?

A Constant temperature   C Constant pressure

B Increasing temperature   D Decreasing pressure [2]

Total Marks _____ / 8

Equilibria

1 Heating a mixture of hydrogen and iodine gas is a reversible reaction:

$$H_2(g) + I_2(g) \rightleftharpoons 2HI(g)$$

When the reaction is carried out in a closed system a dynamic equilibrium will eventually be reached.

Use the following words to complete the sentences below.
Each word may be used once, more than once or not at all.

forward   **reverse**   **H$_2$**   **I$_2$**   **HI**

At the beginning of the reaction, when the reactants are first mixed, the _____ reaction is

dominant. Over time, the concentration of _____ increases and so the _____ reaction will

increase. Eventually, the rate of the forward reaction will equal the rate of the _____ reaction.   [4]

2 HT Which of the following describes Le Chatelier's principle?

   A When the conditions of a system are kept the same, the position of the equilibrium changes to try to maintain the conditions.

   B When the conditions of a system are altered, the position of the equilibrium changes to try to restore the original conditions.

   C When the conditions of a system are altered, the amount of the reactants changes to try to restore the original conditions.

   D When the conditions of a system are altered, the amount of the products changes to try to restore the original conditions.   [1]

3 HT Look at the following reaction:

$$N_2(g) + 3H_2(g) \rightleftharpoons 2NH_3(g)$$

The reaction is exothermic in the forward direction (reactants → product).

Which of the following will lead to an **increase** in the amount of product produced?

   A Reducing pressure

   B Increasing pressure

   C Reducing temperature

   D Increasing temperature   [2]

4 HT A forward reaction is endothermic. The temperature is increased.

Would this change lead to more reactants or more products being produced?   [1]

Total Marks _____ / 8

Practice Questions

Improving Processes and Products

1 Which of the following metals **cannot** be displaced by carbon?

    A Iron   **B** Magnesium   **C** Copper   **D** Lead   [1]

2 Zac is extracting copper using carbon displacement.
He sets up the equipment as shown below:

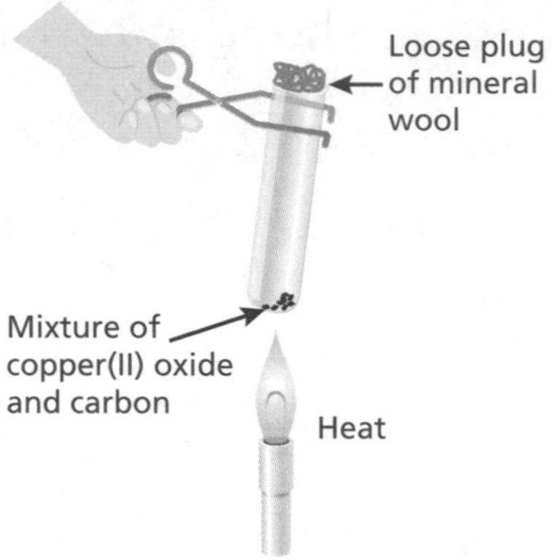

Loose plug of mineral wool

Mixture of copper(II) oxide and carbon

Heat

    a) Write the word equation for the displacement of copper
from copper(II) oxide by carbon. [1]

    b) Suggest why there is a loose plug of mineral wool in the
top of the tube. [1]

3 Which of the following is used to extract metals that are more
reactive than carbon from their ores?

    A Magnetism   **B** Chromatography   **C** Explosives   **D** Electrolysis   [1]

4 For an ionic compound, such as sodium chloride, to be electrolysed, which of the following
statements are true?

    A The amount of metal in the compound has to be greater than 90%.

    B The ionic compound needs to be molten or in solution.

    C A catalyst needs to be added to the ionic compound.

    D The ionic compound needs to be in a gaseous state. [1]

5 **HT** Write the balanced symbol equation for the reaction that occurs at the cathode during
the electrolysis of copper(II) sulfate solution. [1]

6 **HT** Molten aluminium oxide is being electrolysed.

    a) What ions are present in aluminium oxide? [1]

    b) Write the balanced symbol equation for the reaction that takes place at:

        i) The cathode. [1]

        ii) The anode. [1]

7 Give **three** advantages of using bacteria to extract metals from contaminated soils. [3]

The Haber Process

1 The Haber process is used to produce ammonia, which is used to make nitrogen fertilisers.

   a) Why is nitrogen gas not used as a fertiliser? [2]

   b) Write the word equation for the reaction that takes place during the Haber process. [1]

2 HT The effect of increasing pressure and temperature on the yield of the Haber process is shown in the graph on the right.

   Explain why the process **is not** carried out at 400atm and 350°C.

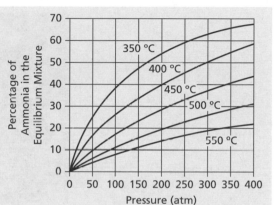

[2]

3 HT Give **three** factors that influence the cost of making a new chemical substance. For each factor, you must explain how it can affect the cost. [3]

4 Producing a chemical in a laboratory is different to the large-scale industrial production of the same chemical.

Two differences between laboratory and industrial scale production are given in the table.

Laboratory Production	Industrial Production
Small quantities of chemicals used	Large quantities of chemicals used
Chemicals are expensive	Chemicals are cheaper

Give **two** other differences between the methods. [2]

Life Cycle Assessments, Recycling and Alloys

1 Which of the following factors are analysed as part of a life cycle assessment?

A Colour   **D** Manufacture   **G** Recycling

B Disposal   **E** Popularity   **H** Transport

C History   **F** Price   **I** Use   [1]

2 Each year thousands of mobile phones are disposed of in landfill sites.
Many of the materials used in the manufacture of mobile phones could be recycled.

a) What is meant by the term **recycled**? [1]

b) Explain why it is important to recycle materials. [4]

c) Suggest why some of the parts of a mobile phone will not be recycled, even though technically it may be possible to do so. [1]

3 The following factors are considered when making a disposable plastic bag.

Which factors would **not** be suitable for inclusion in a life cycle assessment?

A Which bags customers prefer to use

B The environmental impact of disposing of the bags

C The energy required to dispose of the bags

D The energy required to make the bags from plant fibres

E How much should be charged for the bag [2]

Total Marks _____ / 9

Using Materials

1 The corrosion of iron is called rusting.
Rusting causes millions of pounds worth of damage each year.

a) Which of the following equations shows the rusting of iron?

A iron + oxygen → iron oxide

B iron + water → hydrated iron(II) oxide

C iron + water + oxygen → hydrated steel oxide

D iron + water + oxygen → hydrated iron(III) oxide [1]

b) There are a number of ways to prevent corrosion.

Draw **one** line from each object to the most suitable form of protection.

iron bridge		oil
iron bench		paint
iron roof		galvanising
steel bicycle chain		plastic coating

[3]

2 Why are glass bottles used for the storage of corrosive chemicals? [1]

3 The Bugatti Veryon is one of the fastest cars
that can be driven legally on the road.
The outer body is made of a carbon fibre
composite.

a) What is meant by the term **composite**? [1]

b) Suggest **two** reasons why carbon fibre was used to make the car's body instead of
traditional steel. [2]

Total Marks _____ / 8

Practice Questions

Organic Chemistry

1 The diagrams below all show some **organic compounds**:

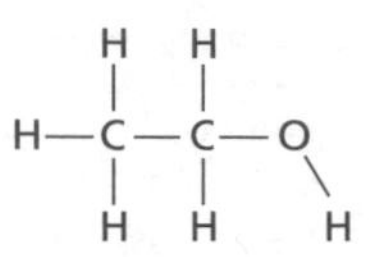

Compound A

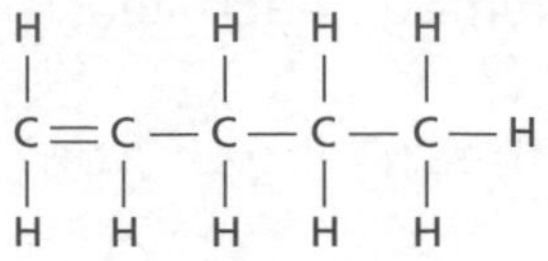

Compound B

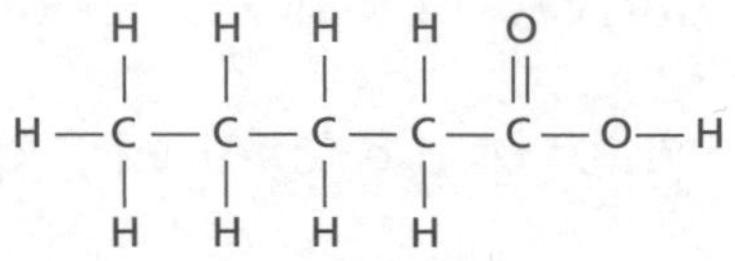

Compound C

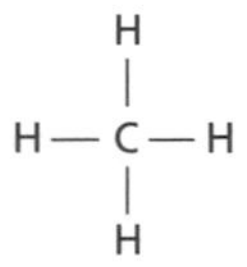

Compound D

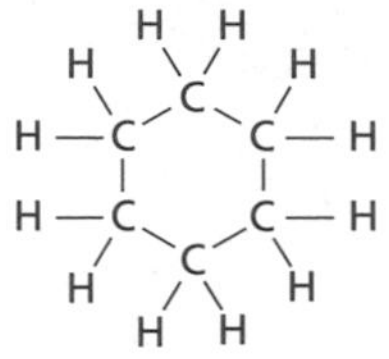

Compound E

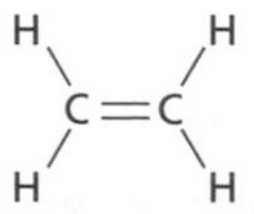

Compound F

    a) Which **two** molecules are in the alkene homologous series? [1]

    b) Which functional group identifies an alkene? [1]

    c) Which **two** molecules are **not** hydrocarbons? [1]

    d) Which compound is a carboxylic acid? [1]

    e) Draw the structural formula for **ethane**. [1]

    f) What is the name of compound **A**? [1]

2 Explain why longer-chain hydrocarbons produce soot when burned. [3]

3 Potassium manganate(VII) is a dark purple colour.
Mary has been given a solution. She suspects it contains an alcohol.
She adds half a spatula of potassium manganate(VII) to the solution.

    a) Write the **word equation** to show the reaction that will take place if an alcohol
is present. [1]

    b) What will Mary observe if an alcohol is present? [1]

Total Marks _____ / 11

Organic Compounds and Polymers

1 DNA is the nucleic acid that codes for our characteristics.
   A diagram of a section of DNA is shown below:

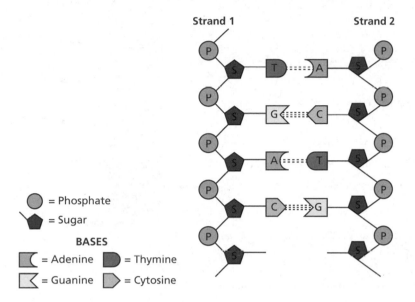

= Phosphate

= Sugar

BASES

= Adenine   = Thymine

= Guanine   = Cytosine

a) DNA is a biological polymer.

    What is the repeating unit (the monomer) in DNA called?   [1]

b) What are the components that make up each repeating unit?   [1]

c) Draw one of the repeating units of DNA.   [1]

2 HT A condensation polymer is made from two monomers.
   One of the monomers has two –COOH groups in its molecule.
   The other monomer has two –NH₂ groups in its molecule.

   Which of the following is the polymer?

   A DNA

   B Poly(ethene)

   C Nylon

   D Polyester   [1]

Total Marks _____ / 4

Crude Oil and Fuel Cells

1. Fractional distillation is used to separate the fractions in crude oil.

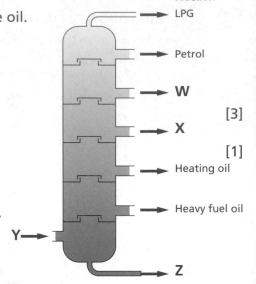

Fraction

→ LPG

→ Petrol

→ **W**

→ **X**

→ Heating oil

→ Heavy fuel oil

Y →

→ **Z**

   a) Choose words from the list below to label **W**, **X**, **Y** and **Z** on the diagram:

   Bitumen   **Crude oil**   **Diesel**   **Paraffin**   [3]

   b) Where is the **lowest temperature** in the column?   [1]

2. These fractions obtained from fractional distillation can be cracked and used elsewhere to make other chemical products.

   Look at the table:

Fraction	Supply (millions barrels per day)	Demand (millions barrels per day)
Petrol	15	40
Paraffin	15	10
Diesel	25	30
Fuel oil	20	10

   Cracking large molecules creates smaller, more useful molecules.

   Which **two** fractions in the table should be cracked to ensure demand is met?   [2]

   Total Marks _____ / 6

Interpreting and Interacting with Earth's Systems

1. Look at the graph showing the relationship between temperature and carbon dioxide levels.

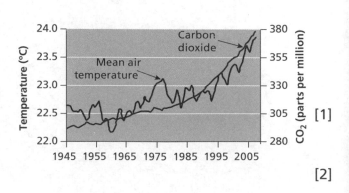

   a) Describe the general trend in air temperature in relation to carbon dioxide levels.   [1]

   b) Explain why carbon dioxide concentrations are linked to temperature increases.   [2]

c) The majority of the world's scientists agree that the causes of climate change are due to the actions of humans.

Which of the following are man-made causes of carbon dioxide emissions?

A   Flying aeroplanes   C   Planting trees

B   Driving cars   D   Burning plastics   [3]

2   The table below shows the gases present in the air.

Gas	Percentage in Clean Air
Nitrogen	78
Oxygen	
Carbon dioxide	0.04
Other gases	1

a) Work out the percentage of oxygen in air.   [1]

b) Describe how the process of **photosynthesis** alters the amounts of carbon dioxide and oxygen in the air.   [2]

c) What were the two main gases present in the Earth's early atmosphere?   [2]

Total Marks .... / 11

Air Pollution, Potable Water and Fertilisers

1   a) Explain why carbon monoxide is toxic in mammals.   [2]

b) What is the main source of carbon monoxide?   [1]

2   Explain why oxides of nitrogen have a negative environmental impact.   [2]

3   Which of the following is the correct explanation for the term **potable water**?

A   Ultra-pure water made from the process of reverse osmosis.

B   Water that is stored in pots and taken from village to village.

C   Water that is safe to drink.

D   Water that needs to be sterilised.   [1]

Total Marks .... / 6

Review Questions

Improving Processes and Products

1 Which of the following metals cannot be extracted from its ore using carbon?

A lithium

B iron

C copper

D lead [1]

2 The industrial electrolysis of aluminium ore takes place in large electrolyte cells like the one shown below.

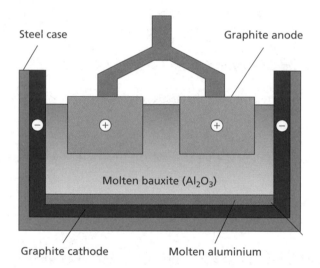

Steel case

Graphite anode

Molten bauxite (Al_2O_3)

Graphite cathode

Molten aluminium

a) Which part of the cell will have to be replaced regularly? [1]

b) HT Which of the following shows the correct ionic equations for the reactions that take place at the electrodes?

Cathode:	Anode:
A $Al^{3+}(l) + 3e^- \rightarrow Al(l)$	$C(s) + O_2(g) \rightarrow CO_2(g)$
B $C(s) + O_2(g) \rightarrow CO_2(g)$	$Al^{3+}(l) + 3e^- \rightarrow Al(l)$
C $Al^{3+}(l) + 3e^- \rightarrow Al(l)$	$2O^{2-}(aq) \rightarrow O_2(g) + 4e^-$
D $2O^{2-}(aq) \rightarrow O_2(g) + 4e^-$	$Al^{3+}(l) + 3e^- \rightarrow Al(l)$

[1]

Total Marks .... / 3

The Haber Process

1 HT Phytoextraction is a way of removing harmful chemicals from the soil.

Which of the following is **not** an advantage of phytoextraction?

A Extracted metal can be reclaimed from the ash of the burned plants.

B It is a cost-effective way of treating soil.

C Plants take a long time to grow.

D Plants can be bred to extract certain heavy metals. [1]

2 The Haber process is used to produce ammonia.

What are the **two** reactants in the Haber process?

A Ammonia

B Hydrogen

C Iron catalyst

D Nitrogen [2]

3 HT The industrial production of chemicals always involves compromises.

a) Describe the typical conditions used for carrying out the Haber process. [3]

b) Explain why these conditions are used instead of the optimum conditions, which would produce a higher yield. [2]

4 HT In industry, it is important that chemicals are produced as cheaply as possible so that they can be sold for a profit.

Give **three** factors that affect the cost of making a new chemical. [3]

Total Marks _____ / 11

Review Questions

Life Cycle Assessments, Recycling and Alloys

1 A life cycle assessment can be divided into four stages.
The four stages are listed below, but they are in the wrong order:

A Disposal of the product.

B Obtaining raw materials / producing the materials needed for the product.

C Use of the product.

D Making the product.

a) List the stages in the correct order. [1]

The table below shows the results of life cycle assessments for three different types of bottle stops: corks, synthetic (plastic) corks and metal screw caps.

	Totals for 1000 Units		
	Cork	**Synthetic Cork**	**Metal Cap**
Energy Use (MJ)	1250	4500	5600
Fossil Fuel Use (kg)	27	54	149
Waste	46	1	90
Greenhouse Gas Emissions (kg CO2)	40	81	231
Fresh Water Use (litres)	8400	4010	4200

b) Give **two** ways in which the manufacture of synthetic corks is **more** environmentally friendly than the manufacture of corks and metal caps. [2]

c) The metal used to make the caps is aluminium.

Suggest why using aluminium results in the highest CO_2 emissions. [2]

d) Which type of cap uses the least energy? [1]

e) What major advantage do aluminium caps have over the other two types? [1]

Total Marks _____ / 7

Using Materials

1 Some iron objects are covered in tin to stop them from rusting.

Give **two** other methods that are used to stop rusting. [2]

2 Composite materials have been used for centuries.

The Romans were amongst the first cultures to use concrete for building purposes. Many of the structures they built using concrete still stand today.

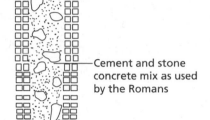
Cement and stone concrete mix as used by the Romans

Explain why the concrete used by the Romans is a **composite material**. [2]

Total Marks .... / 4

Organic Chemistry

1 Look at the displayed formulae of the following compounds:

A

B

C

D

E

F

a) Which of the compounds are **saturated**?
   You must explain your answer. [2]

b) Which of the compounds are **hydrocarbons**? [1]

c) Which of the compounds are **alkenes**? [1]

d) What type of compound is compound **D**? [1]

e) What are the functional groups of compound **D**? [1]

Review Questions

2 Hydrocarbons can produce soot when they are burned.

Which of the following hydrocarbons is likely to produce the most soot?

A C_2H_6

B $C_{10}H_{22}$

C C_2H_4

D C_8H_{18} [1]

Total Marks _____ / 7

Organic Compounds and Polymers

Poly(propene)

1 Poly(propene) is a polymer that is used to make rope.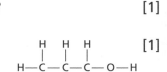

a) Draw the **monomer** unit for poly(propene). [1]

b) What is the main difference between poly(propene) and poly(ethene)? [1]

2 a) What is the chemical name of **X**? [1]

b) Compound **X** is heated with acidified potassium manganate(VII) solution.
The product of the reaction is compound **Y**.

i) What is the name of compound **Y**? [1]

ii) Describe what happens to the colour of the acidified potassium manganate(VII) solution when compound **Y** is produced. [1]

3 A student bubbles ethene gas into bromine water.

Draw the molecule that is formed by this reaction. [1]

Ethene

Total Marks _____ / 6

Crude Oil and Fuel Cells

1 Teflon is the trade name of a chemical that is used to make non-stick surfaces for cookware. The scientific name for Teflon is poly(tetrafluoroethene). Teflon is an addition polymer.

$$\left(\begin{matrix} F & F \\ | & | \\ -C & -C- \\ | & | \\ F & F \end{matrix}\right)_n$$

a) Draw the monomer unit for Teflon. [1]

b) One of the long-chain alkanes that is cracked is $C_{16}H_{34}$.
One mole of cracked $C_{16}H_{34}$ leads to the formation of one mole of C_2H_4, plus a second product.

What is the formula of the second product? [1]

c) Write a balanced symbol equation for this reaction. [1]

2 The following are fractions of crude oil:

bitumen   **petrol**   **diesel**   **paraffin**

Which fraction has the **lowest** boiling point? [1]

Total Marks _____ / 4

Review Questions

Interpreting and Interacting with Earth's Systems

1 Carbon dioxide levels are increasing globally.

The Kyoto agreement was an international agreement, made in 2012, to reduce CO_2 levels to a certain level.

One way to capture carbon dioxide is to inject it into wells, which are between 800 and 3300m deep, to store the greenhouse gas.

The table below shows:

- the estimated number of wells that are now needed to keep CO_2 at the 2015 levels
- the number of wells predicted by the Kyoto agreement in 2012.

Year	Wells Now Needed to Meet Kyoto Target Levels	Wells Predicted in 2012
2015	100 176	40 332
2020	120 342	60 498
2025	140 508	80 664
2030	160 674	100 830

a) Explain why it is important to reduce the CO_2 levels in the atmosphere. [3]

b) Suggest why the number of wells needed to meet the Kyoto target now is greater than was predicted. [2]

c) Calculate the percentage increase in the wells now needed to meet Kyoto target levels from 2015 to 2030. [1]

d) Scientists believe that human activity has caused climate change.

Give **two** ways in which humans are believed to have increased the amount of CO_2 in the atmosphere. [2]

2 These statements explain how scientists think that our modern-day atmosphere evolved.

1. Nitrogen gas was released from ammonia by bacteria in the soil.

2. The modern atmosphere consists of nitrogen, oxygen and a very small amount of carbon dioxide.

3. Plants evolved and used carbon dioxide for photosynthesis, producing oxygen as a by-product.

4. Volcanoes gave out ammonia, carbon dioxide, methane and water vapour.

5. As the Earth cooled, water fell as rain, which led to the formation of oceans.

   Which is the correct order of this happening?

   A 4, 5, 1, 3, 2

   B 2, 5, 3, 1, 4

   C 1, 4, 3, 5, 2

   D 4, 3, 1, 5, 2 [1]

Total Marks _____ / 9

Air Pollution, Potable Water and Fertilisers

1 Explain why carbon monoxide is poisonous to mammals. [3]

2 Which of the following is **not** a consequence of air pollution from SO_2?

   A Acid rain

   B Phytochemical smog

   C Fine particulates

   D Lung disease [1]

3 Give **two** chemical methods for making drinking water from sea water. [2]

4 The diagram shows a form of reverse osmosis.

   a) Name the part of the diagram labelled **X**. [1]

   b) Suggest where this method of producing water is
   most likely to be needed. [2]

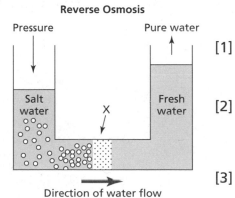

Reverse Osmosis

Pressure · Pure water · Salt water · X · Fresh water · Direction of water flow

5 Fertilisers are used on farmland to replenish natural minerals.

   a) What are the **three** main elements used in fertiliser? [3]

   b) Suggest how ammonium phosphate can be made using neutralisation. [1]

Total Marks _____ / 13

Mixed Exam-Style Questions

1 Christine is investigating the reaction between calcium carbonate and dilute nitric acid.
She measures the reaction time for four different concentrations of nitric acid.
The only factor that was changed was concentration.
The results are shown in the table below.

Concentration	Reaction Time (s)
A	32
B	63
C	97
D	144

Explain, in terms of particles, why **A**, **B**, **C** and **D** had different reaction times. [4]

2 Eliot is carrying out chromatography on inks from different pens.
He is trying to find out which ink was used to write a note.
He draws a start line in pencil and then puts a dot of each ink onto the line.
He then dips the filter paper into ethanol in a beaker.

a) What is the role of the **paper** in this experiment?

    A Mobile phase   **C** Solid phase

    B Gas phase   **D** Stationary phase [1]

b) Why did Eliot draw the start line in pencil? [1]

c) The results of Eliot's experiment are shown below.

    Which ink matches sample **X**, which was used on the note? [1]

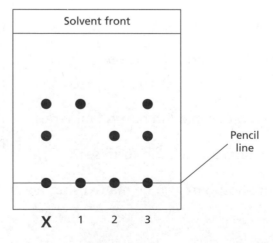

d) Calculate the R_f value for the ink in sample 2. Use a ruler to help you. [1]

3 HT During the electrolysis of molten lithium chloride, what is made at the anode?

   A Chlorine

   B Hydrogen

   C Lithium

   D Lithium hydroxide   [1]

4 The diagrams below show the structures of two forms of carbon:

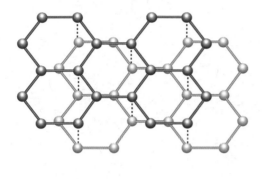 

a) Diamond is very hard material.
Graphite is not very hard.

   Use ideas about bonding and structure to explain this observation.   [4]

b) What is the technical term given to the different forms carbon can take?   [1]

c) Describe **one** way of demonstrating that graphite and diamond are different physical forms of carbon.   [2]

5 Ethane has the formula C_2H_6.
Look at the three different representations of ethane.

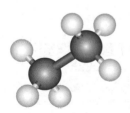

Ball and stick model   Displayed formula   Dot and cross diagram

a) What type of bonding is present in ethane?
You must explain how the diagrams show this.   [2]

b) Describe the limitations of using displayed formulae to represent molecules.   [2]

6 Lithium is extracted from its ore by electrolysis.
Iron is extracted from its ore by heating it with carbon.

a) Explain why different methods are used to extract lithium and iron. [2]

b) Molten lithium oxide contains Li^+ and O^{2-} ions.
The electrolysis of molten lithium oxide produces lithium and oxygen.

   i) Write the **balanced symbol** equation for the electrode reaction that happens at the cathode.
   Use e^- to represent an electron. [2]

   ii) Explain why solid lithium oxide cannot be electrolysed. [1]

7 HT Magnesium carbonate reacts with sulfuric acid to form carbon dioxide, magnesium sulfate and water.
A graph of the reaction is shown. [1]

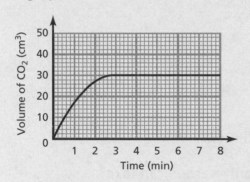

What is the rate of reaction between 0min and 2min?

8 Ethene has the formula C_2H_4.
Ethene is an **unsaturated** compound.

Explain what is meant by the term **unsaturated**. [1]

9 The table contains information about some atoms and ions.

Particle	Atomic Number	Atomic Mass	Number of Protons	Number of Neutrons	Number of Electrons	Electronic Structure
A	7	14	7		7	2.5
B		40	20	20	18	2.8.8
C	8		8	8	8	
D	9	19		10	9	2.7

a) Complete the table. [5]

b) Explain why particle **D** is an atom and particle **B** is an ion. [2]

c) Particle A has the electronic structure 2.5.

What does this tell you about the position of particle **A** in the periodic table? Explain your answer. [4]

d) Another particle has an atomic mass of 40 and contains 21 neutrons.

What is the atomic number of particle **B**? [1]

10 Which technique would be the best for separating pure water from sea water?

A Chromatography

B Filtration

C Distillation

D Crystallisation [1]

11 Two isotopes of carbon are:

$^{12}_{6}C$ $^{14}_{6}C$

Explain the difference between isotopes of the same element in terms of subatomic particles? [4]

12 The table below shows some fractions of crude oil and the range of their boiling points.

Fraction	Boiling Point Range (°C)
Petrol	80–110
Diesel	150–300
Fuel oil	300–380
Bitumen	400+

A hydrocarbon called hexadecane has a boiling point of 289°C.

In which fraction would you expect to find hexadecane? [1]

13 Devlin reacts 0.9g of element **X** with 7.1g of chlorine, Cl_2.
There is one product: **X** chloride.

a) What mass of **X** chloride is produced? [1]

b) Calculate the number of moles of **X**, chlorine (Cl_2) and **X** chloride involved in the reaction.
(The relative atomic mass of **X** = 9. The relative formula mass of chlorine = 71 and
X chloride = 80.) [1]

c) Use your answers to construct the **balanced symbol** equation for this reaction. [2]

14 Ras Al-Khair in Saudi Arabia produces around one million cubic litres of drinking water a day
from sea water.

Which of the following describes the process of extracting fresh water from sea water?

A Chlorination

B Filtration

C Desalination

D Sedimentation [1]

15 Magnesium sulfate is a salt.
It can be made by reacting magnesium oxide, MgO, with sulfuric acid, H_2SO_4.

a) Construct the **balanced symbol** equation for this reaction. [1]

b) Naomi suggests another method for preparing magnesium sulfate:

1. Measure 50cm³ of dilute sulfuric acid into a beaker.
2. Add one spatula of magnesium oxide.
3. Heat the mixture until only crystals of magnesium sulfate are left.

Give **one** safety precaution that Naomi should take when following this method. [1]

c) Naomi's method will **not** make a pure, dry sample of magnesium sulfate.

How can Naomi make sure that:

i) The reaction is complete? [1]

ii) The filtrate contains only magnesium sulfate solution? [1]

16 Look at the diagram of four different atoms:

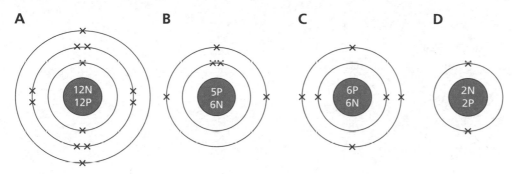

a) What is the centre of the atom called? [1]

b) What subatomic particles make up the centre of the atom? [1]

c) Which two elements shown are in the same **period** in the periodic table? [1]

d) What are the names of elements **B** and **C**? [2]

17 The formula for butanoic acid is $CH_3CH_2CH_2COOH$.

a) What is the **empirical formula** of butanoic acid? [1]

b) Calculate the relative formula mass for butanoic acid.
(The relative atomic mass of C = 12, H = 1 and O = 16.) [1]

18 Titanium is a metal.

Which statement is true about titanium **because** it is a metal?

A It is a shiny silver colour

B It is malleable

C It is in Period 4 of the periodic table

D It is corrosion resistant [1]

19 Look at the diagrams of compounds below.

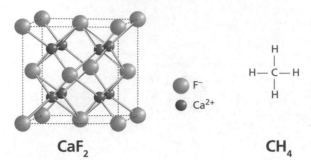

CaF$_2$ CH$_4$

a) Explain why the structure of calcium fluoride, CaF$_2$, means it has a high melting point. [3]

b) Explain why methane, CH$_4$, has a low melting point and boiling point. [3]

c) Draw dot and cross diagrams to show the ionic bonding in lithium oxide.
You should include the charges on the ions.
The electronic structure of lithium is 2.1.
The electronic structure of oxygen is 2.6. [2]

20 a) What is meant by the term **pure substance** in chemistry? [1]

b) Explain how this is different to the way in which pure is used to describe substances, e.g. orange juice, in everyday life. [1]

21 Look at the elements from the periodic table and answer the questions that follow.

$^{9}_{4}Be$ $^{24}_{12}Mg$ $^{27}_{13}Al$ $^{19}_{9}F$ $^{31}_{15}P$

a) Which element has a relative atomic mass of 9? [1]

b) Which element has an atomic number of 9? [1]

c) An isotope of P has two fewer neutrons.

What is the relative atomic mass of the isotope? [1]

22 HT Crude oil is separated into different fractions during fractional distillation.
Each fraction contains hydrocarbon molecules that have similar boiling points.
The boiling point of a hydrocarbon is determined by the number of carbon atoms in it.

	Boiling Point	No. of Carbon Atoms	Size of Intermolecular Forces
A	Low	More than 50	Large
B	High	Less than 20	Large
C	Low	Less than 20	Small
D	High	More than 50	Small

Which letter, **A**, **B**, **C** or **D**, represents the **correct** relationship between boiling point,
number of carbon atoms and the size of the intermolecular forces in a molecule? [1]

23 Here is the displayed formula of an organic compound:

What is the name of this compound?

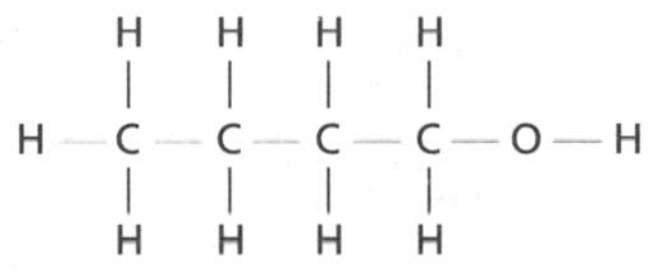

A Butanoic acid

B Butanol

C Propanoic acid

D Propanol [1]

24 Magnesium and dilute sulfuric acid react to make hydrogen gas:

$Mg(s) + H_2SO_4(aq) \rightarrow MgSO_4(aq) + H_2(g)$

George measures the rate of this reaction by measuring the **loss in mass** of the reaction mixture.
He finds that the change in mass is very small and difficult to measure.

a) Draw a labelled diagram to show a better way of measuring the rate of this reaction. [1]

b) The reaction between the magnesium and sulfuric acid takes longer than George would like.

What type of substance can George add to increase the rate of reaction? [1]

c) Suggest what safety precautions George needs to take when carrying out the experiment. [1]

25 **a)** The table contain some information about different polymers.

Polymer	Density (g/dm³)	Melting Point (°C)	Solubility in Oil
A	0.92	85	Insoluble below 80°C; soluble above 80°C
B	0.96	120	Insoluble below 80°C; soluble above 80°C
C	1.05	65	Soluble
D	1.39	60	Soluble
E	0.90	150	Insoluble

A manufacturer needs to use a polymer to make a pipe which will carry oil at a temperature of 100°C.

Explain which polymer they should use. [3]

b) What is the name given to the small molecules that join together to form a polymer? [1]

c) Look at the displayed formula of chloroethene.
Chloroethene is used to make a polymer.

Draw the displayed formula of the polymer showing three repeat units. [2]

26 Jess tests an unknown compound using the flame test.
The unknown compound contains both calcium ions and lithium ions.

a) Suggest why Jess might find it difficult to interpret the results of the flame test. [1]

b) Jess carries out three chemical tests on an unknown solution.
Her results are shown in the table.

Chemical Test	Result
pH probe	4
Barium chloride solution followed by dilute hydrochloric acid	Precipitate disappears
Silver nitrate solution followed by dilute nitric acid	Cream precipitate

Which ions are present in the solution?

Choose from:

iron(II)   bromide   hydrogen   hydroxide   carbonate   calcium

Explain your answer. [3]

27 Iodine is a non-metal.

Which statement is true about iodine **because** it is a non-metal?

A It is a solid at room temperature and pressure.

B It does not conduct electricity.

C It is a halogen.

D It is in Period 5 of the periodic table.   [1]

28 HT Which of the following shows the approximate size of an atom?

A 3×10^3m

B 3×10^{-3}m

C 3×10^{13}m
                                                               [1]
D 3×10^{-23}m

29 Karen adds sodium hydroxide solution to a sample of calcium chloride solution.

What is the colour of the precipitate formed?

A Blue

B Green

C Orange

D White   [1]

30 Baking powder is added to a cake mixture.
Baking powder contains sodium hydrogen carbonate, $NaHCO_3$.
When sodium hydrogen carbonate is heated, it makes sodium carbonate, Na_2CO_3, carbon dioxide and water.

a) Write the **balanced symbol** equation for this reaction.   [2]

b) What is the charge on a single hydrogen carbonate ion?   [1]

c) Describe the chemical test to detect carbonate ions in solution.   [2]

Mixed Exam-Style Questions

31 Sea water contains many useful chemicals.
A company in the Middle East, specialising in desalination of sea water, decides that they can also extract particular chemicals and sell them onto the chemical industry.

a) What is desalination? [1]

b) Sodium chloride and sodium sulfate can both be extracted from sea water.

What is the chemical formula for sodium sulfate? [1]

c) A technician tests sea water with dilute nitric acid followed by silver nitrate solution.
A white precipitate is formed.

Suggest what conclusion can be drawn from this result. [1]

d) It is important that people in all parts of the world have access to potable water.
Explain why. [2]

32 Drema Chemicals manufacture chemicals for use in the chemical industry.
They make hydrogen peroxide, H_2O_2, to be used as a rocket propellant.
Hydrogen peroxide can be manufactured using the two methods shown in the table:

	Method 1	Method 2
Reactants	Barium peroxide and sulfuric acid	Hydrogen and oxygen
Temperature	5°C	45°C
Catalyst used?	no	yes
Percentage yield	70%	95%
Pollution problems	Poisonous waste product made	No waste products made

Drema Chemicals decide to use Method 2 because it is cheaper.

Using the table, suggest **two** reasons why Method 2 is cheaper than Method 1. [2]

33 Nanoparticles are often used as catalysts.

a) Why do nanoparticles make good catalysts? [1]

b) What are the risks of using nanoparticles? [1]

34 Ethanol has the formula C_2H_5OH.

   a) How many atoms of C, H and O are present in one molecule of ethanol? [1]

   b) Ethene is an alkene.

      Draw the **displayed formula** of ethene. [1]

   c) Draw the **displayed formula** of ethanol. [1]

   d) What is the **functional group** of ethanol? [1]

35 What is the general formula for the reaction between a metal and an acid? [1]

36 Complete the word equations to show the products of each reaction:

   a) magnesium + nitric acid [1]

   b) zinc + sulfuric acid [1]

   c) iron + hydrochloric acid [1]

37 What is the general formula for the reaction between metal carbonates and acid? [1]

38 Complete the word equations to show the products of each reaction:

   a) magnesium carbonate + nitric acid [1]

   b) copper carbonate + sulfuric acid [1]

   c) sodium carbonate + hydrochloric acid [1]

Total Marks _____ / 119

Answers

Pages 6–7 KS3 Review Questions

1. Distilled water = compound [1]; Gold = element [1]; Glucose = compound [1]; Salt water = mixture [1]
2. a) B [1]; C [1]
   b) magnesium + oxygen → magnesium oxide [2] (1 mark for correct reactants; 1 mark for correct product)
3. A correctly drawn diagram of the particles in a solid [1]; liquid [1]; and gas [1]

Solid   Liquid   Gas

4. D [1]
5. a) An alkali / base [1]
   b) B [1]
6. a) The elements are very unreactive [1]; and will not react with oxygen or other elements [1]
   b) **Any four from**: Aluminium is very reactive [1]; aluminium is higher than carbon in the reactivity series [1]; aluminium can only be extracted using electricity / electrolysis [1]; electrictiy was not readily available until the beginning of the 19th century [1]; gold and platinum would be found in elemental / unreacted form [1]
7. a) Red [1]
   b) Yellow [1]
   c) Green [1]
   d) Blue [1]
8. a) D [1]
   b) C [1]
   c) B [1]
9. Solid water / ice is less dense than liquid water [1]; this means that under the ice there is liquid water [1]

Pages 8–17 Revise Questions

Page 9 Quick Test
1. Different forms of an element with the same number of protons but a different number of neutrons
2. 8 (14 – 6 = 8)
3. New experimental evidence has led to changes

Page 11 Quick Test
1. 58.3 g
2. Mixtures of substances in solution
3. CH_3

Page 13 Quick Test
1. By atomic number; by number of electrons in the outermost shell
2. Magnesium, 2.8.2

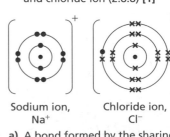

3. 2.8.7

Page 15 Quick Test
1. The electrons in the outermost shells of the metal atoms are free to move, so there are a large number of electrons moving between the metal ions.
2. **Any two from:** Distances between electrons and the nucleus are not realistic; bonds appear to be physical structures; bond lengths are not in proportion to the size of the atom; they do not give a good idea of the 3D shape of the atoms.
3. A 3D arrangement of a large number of repeating units (molecules / atoms) joined together by covalent bonds.

Page 17 Quick Test
1. A compound that contains carbon
2. It is a giant covalent molecule in which every carbon atom forms bonds with four other carbon atoms (the maximum number of bonds possible).
3. 1–100nm

Pages 18–21 Practice Questions

Page 18 Particle Model and Atomic Structure
1. a) D [1]
   b) The temperature is increasing [1]; so the particles in the solid ice start vibrating more quickly [1]
2. There is nothing in the gaps / empty space between the water molecules [1]
3. a) Ne [1]
   b) Ca [1]
   c) 28 + 2 = 30 [1]

> It is usual for mass numbers to be rounded to the nearest whole number.

4. a) A heating curve drawn showing a straight line at 100°C [1]; and a straight line at 0°C [1]; sloping down in direction left to right [1]; with correctly labelled axes [1]

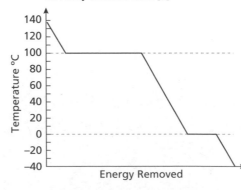

   b) The water molecules slow down [1]; as they lose kinetic energy [1]
5. a) charge = – / negative [1]; relative mass = 0.0005 / zero / negligible [1]
   b) charge = + / positive [1]; relative mass = 1 [1]
   c) charge = 0 / neutral [1]; relative mass = 1 [1]

Page 19 Purity and Separating Mixtures
1. a) Containing one type of atom or molecule only [1]
   b) Every substance has a specific melting point at room temperature and pressure [1]; if the substance melts at a different temperature, it indicates that there are impurities [1]
2. a) Distance moved by the solvent = 28 [1]; R_f = $\frac{\text{distance moved by the compound}}{\text{distance moved by the solvent}}$,

   R_f (pink) = $\frac{7.5}{28}$ = 0.27 [1];

   R_f (purple) = $\frac{17.5}{28}$ = 0.63 [1]

   b) D [1]
3. C_4H_4S [1]
4. C = $\frac{84}{12}$ = 7 [1]; H = $\frac{16}{1}$ = 16, O = $\frac{64}{16}$ = 4 [1]; $C_7H_{16}O_4$ [1]

> Look for common factors to see whether an empirical formula can be simplified further.

5. A [1]

Page 19 Bonding
1. a) H = 1, Na = 11, Mg = 12, C = 6, O = 8, K = 19, Ca = 20, Al = 13 [2] (1 mark for 6–7 correct; 0 marks for 5 or less correct)
   b) (2 × 39.1) + 16 = 94.1 [1] (Accept 94)
   c) 16 × 2 = 32 [1]
2. a) They all have the same number of electrons in their outer shell [1]
   b) They all have the same number of shells / they all have two shells [1]
3. a) C [1]
   b) B = 2.8.1 [1]; E = 2.8.7 [1]
   c) Electrons [1]

Page 20 Models of Bonding
1. a) An atom or molecule that has gained or lost electrons [1]
   b) A correctly drawn sodium ion (2.8) [1]; and chloride ion (2.8.8) [1]

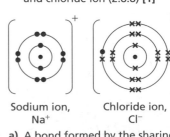

Sodium ion,   Chloride ion,
Na⁺   Cl⁻

2. a) A bond formed by the sharing [1]; of two outer electrons [1]
   b) Two correctly drawn chlorine atoms (each with 7 electrons) [1]; overlapping and sharing two electrons [1]

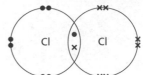

3. a) A large structure that is made up of repeating units (atoms / molecules) [1]; that are covalently bonded together [1]
    b) The atoms in the water / H_2O molecules are covalently bonded [1]; but each water molecule is separate – there is no covalent bond between the water molecules [1]

Page 21 Properties of Materials

1. a) **Any three from:** Carbon has four electrons in its outer shell [1]; it can form covalent bonds [1]; with up to four other atoms [1]; and can form chains [1]
    b) An allotrope is a different form of an element [1]
    c) **Any three from:** graphite [1]; diamond [1]; fullerene / buckminsterfullerene [1]; graphene [1]; lonsdaleite [1]; amorphous carbon [1]
2. a) It conducts electricity because it has free electrons [1]; and it is stronger than steel [1]
    b) Diamond does not conduct electricity [1]
    c) A diagram showing each carbon joined to four other carbon atoms [1]; with a minimum of five atoms shown in tetrahedral arrangement [1]

Diamond

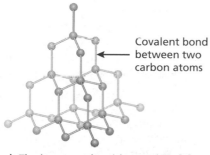

Covalent bond between two carbon atoms

3. a) The ions must be either molten [1]; or dissolved in aqueous solution [1]
    b) In crystalline form, the distance between the ions is at its smallest / the ions are close together [1]; so the electrostatic forces are very high and have to be overcome for the crystal to melt [1]
4. C [1]

Pages 22–35 Revise Questions

Page 23 Quick Test

1. $Ca(OH)_2$
2. $2Na(s) + Cl_2(g) \rightarrow 2NaCl(s)$
3. $2Mg, 2S, 8O$ ($2 \times O_4$)

Page 25 Quick Test

1. $BaO, CuF_2, AlCl_3$
2. Al_2O_3

---

3. $Ba^{2+}(aq) + CO_3^{2-}(aq) \rightarrow BaCO_3(s)$

> The $BaCO_3$ formed is insoluble.

Page 27 Quick Test

1. mass = number of moles × relative molecular mass
2. 18g
3. 2.2×10^{-22}g

Page 29 Quick Test

1. The minimum amount of energy needed to start a reaction
2.

Reaction Profile for an Endothermic Reaction

Energy

Energy is absorbed from surroundings

Products

Reactants

Progress of Reaction

3. Exothermic

Page 31 Quick Test

1. Carbon dioxide
2. Zinc nitrate
3. copper oxide + sulfuric acid → copper sulfate + water

Page 33 Quick Test

1. Universal indicator can show a range of pHs from 1 to 14. Litmus paper only shows if something is an acid or alkali.
2. The volume of acid needed to neutralise the alkali; the pH when a certain amount of acid has been added.
3. pH is a measure of the number of H^+ ions in solution.

Page 35 Quick Test

1. Magnesium will be formed at the cathode and bromine at the anode.
2. Hydrogen will be formed at the cathode and bromine at the anode.
3. $Al^{3+} + 3e^- \rightarrow Al$

Pages 36–39 Review Questions

Page 36 Particle Model and Atomic Structure

1. a) It has evaporated [1]
    b) Increase the temperature of the room / switch on a fan [1]
    c) i) A diagram showing more than five particles, with at least 50% of the particles touching [1]
        ii) A diagram showing more than five particles, with none of the particles touching [1]

In cup   In air

---

2. D [1]
3.

Subatomic Particle	Relative Mass	Relative Charge
Proton [1]	1	+1 (positive) [1]
Neutron	1 [1]	0 (neutral)
Electron	Negligible [1]	–1 (negative) [1]

4. a) Protons [1]
    b) The positively charged protons are at the centre of the atom [1]; so, only alpha particles travelling close to the centre of the atom will be repelled / deflected by them [1]

> Alpha particles are positively charged. Like charges repel each other.

5. **He:** 2 (4 – 2) [1]; **Na:** 12 (23 – 11) [1]; **V:** 28 (51 – 23) [1]

Page 37 Purity and Separating Mixtures

1. a) Measure its boiling point [1]; compare the boiling point with data from a data book / known values [1]
    b) The water is not pure [1]; it contains other substances [1]
2. a) CH_2O [1]

> In $C_6H_{12}O_6$ the common factor of all the numbers is 6, so divide by six to simplify the formula.

    b) CH_2O [1]

> Collect all the atoms of the same element together first, and then simplify: $CH_3COOH \rightarrow C_2H_4O_2 \rightarrow CH_2O$

    c) C_2H_4O [1]
3. a) $(6 \times 12) + (12 \times 1) + (6 \times 16) = 180$ [1]
    b) $(2 \times 12) + (4 \times 1) + (2 \times 16) = 60$ [1]
    c) $(1 \times 12) + (2 \times 16) = 44$ [1]
    d) $(2 \times 1) + (32.1) + (4 \times 16) = 98.1$ [1] (Accept 98)
4. Alloy [1]
5. C [1]

Page 38 Bonding

1. Mendeleev [1]
2. a) A [1]; E [1]
    b) A [1]; B [1]
    c) B [1]
    d) A = 3 [1]; B = 11 [1]; C = 13 [1]; D = 18 [1]; E = 4 [1]
3. a) **Any three from:** malleable [1]; sonorous [1]; ductile [1]; form a positive cation [1]; shiny [1]
    b) Metal oxide [1]
    c) An ionic compound [1]

Page 38 Models of Bonding

1. C [1]

Answers

2. One carbon atom drawn **[1]**; with covalent bonds with four hydrogen atoms **[1]**

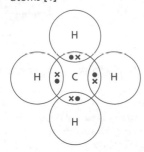

3. Ball and stick models give a better picture of the 3D shape of the molecule **[1]**; and the bond angles / directions **[1]**

4. A correctly drawn magnesium atom, 2.8.2 **[1]**; and magnesium ion $(2.8)^{2+}$ **[1]**

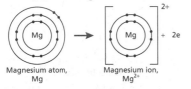

Magnesium atom, Mg

Magnesium ion, Mg^{2+}

Page 39 Properties of Materials

1. a) A different physical structure to other forms of the element **[1]**
   b) Graphite is made of layers of atoms **[1]**; these are held together by weak forces **[1]**; so the layers can separate / slide over each other easily **[1]**; preventing the surfaces from rubbing together **[1]**
   c) In diamond all possible covalent bonds have been used / each carbon atom is bonded to four other carbon atoms **[1]**; so it is extremely hard **[1]**; and has a high melting point **[1]**

   > You need to mention high melting point to get the third mark. Drill bits get hot due to frictional forces, so it is an important property.

   d) **Any one from:** electrical components **[1]**; solar panels **[1]**
2. a) Silver has antibacterial properties **[1]**
   b) They are 1–100nm in size (in the nanometre range) **[1]**
   c) **Any two from:** they can easily get into the human body / cells / the environment **[1]**; they can catalyse reactions **[1]**; they can kill good bacteria **[1]**; the effect on the human immune system is not known **[1]**
3. large surface area **[1]**

Pages 40–45 Practice Questions

Page 40 Introducing Chemical Reactions

1. In a chemical reaction, the mass of the reactants will always equal to the mass of the products **[1]** (Accept: No atoms are made or destroyed)

2. a) The number of atoms of that element present (the element before the number) **[1]**
   b) i) C = 6, H = 12, O = 6 **[1]**
   ii) C = 3, H = 6, O = 2 **[1]**
   iii) H = 2, O = 2 **[1]**
   iv) Ca = 1, N = 2, O = 6 **[1]**
3. solid = (s), liquid = (l), gas = (g), aqueous = (aq) **[1]**
4. a) $2Mg(s) + O_2(g) \rightarrow 2MgO(s)$ **[2]**
   (1 mark for correct balancing; 1 mark for correct state symbols)
   b) $4Li(s) + O_2(g) \rightarrow 2Li_2O(s)$ **[2]**
   (1 mark for correct balancing; 1 mark for correct state symbols)
   c) $CaCO_3(s) + 2HCl(aq) \rightarrow CaCl_2(aq) + CO_2(g) + H_2O(l)$ **[2]** (1 mark for correct balancing; 1 mark for correct state symbols)
   d) $4Al(s) + 3O_2(g) \rightarrow 2Al_2O_3(s)$ **[2]** (1 mark for correct balancing; 1 mark for correct state symbols)

Page 40 Chemical Equations

1. a) 2+ **[1]**
   b) 2– **[1]**
   c) 3+ **[1]**
   d) 2– **[1]**
2. a) $2H^+(aq) + 2e^- \rightarrow H_2(g)$ **[1]**
   b) $Fe^{2+}(aq) + 2e^- \rightarrow Fe(s)$ **[1]**
   c) $Cu^{2+}(aq) + 2e^- \rightarrow Cu(s)$ **[1]**
   d) $Zn(s) \rightarrow Zn^{2+}(aq) + 2e^-$ **[1]**
3. $Ag^+(aq) + Cl^-(aq) \rightarrow AgCl(s)$ **[2]** (1 mark for the correct ions and product; 1 mark for the correct charges)

Page 41 Moles and Mass

1. One mole of a substance contains the same number of particles as the number of atoms in 12g of the element carbon-12 **[1]**
2. B **[1]**
3. g/mol **[1]**
4. a) number of moles of Mo =
   $\frac{mass}{relative\ molecular\ mass}$ =
   $\frac{287.7g}{95.9g/mol}$ **[1]**; = 3mol **[1]**
   b) mass = relative molecular mass × number of moles = 50.9g/mol × 5 mol **[1]**; = 254.5g **[1]** (Accept 255g)

   > Rearrange the equation
   > moles = $\frac{mass}{relative\ molecular\ mass}$
   > to work out the mass.

5. $(6 \times 12) + (12 \times 1) + (6 \times 16) = 180$ **[1]**
6. a) mass of one atom (V) =
   $\frac{atomic\ mass}{Avogadro's\ constant} = \frac{50.9g}{6.022 \times 10^{23}}$ **[1]**; = 8.5×10^{-23}g **[1]**

   b) mass of one atom (Mo) =
   $\frac{95.9g}{6.022 \times 10^{23}}$ **[1]**; = 1.6×10^{-22}g **[1]**

   c) mass of one atom (Cs) =
   $\frac{132.9g}{6.022 \times 10^{23}}$ **[1]**; = 2.2×10^{-22}g **[1]**

   d) mass of one atom (Bi) =
   $\frac{209g}{6.022 \times 10^{23}}$ **[1]**; = 3.5×10^{-22}g **[1]**
7. $2H_2(g) + O_2(g) \rightarrow 2H_2O(l)$,
   1mol of $H_2 \rightarrow$ 1mol of H_2O, so
   5mol of $H_2 \rightarrow$ 5mol of H_2O **[1]**;
   1mol of $H_2O = (2 \times 1) + 16 = 18g$, so
   5mol of $H_2O = 5 \times 18 = 90g$ **[1]**

Page 42 Energetics

1. a) Exothermic **[1]**
   b) Endothermic **[1]**
2. **Any two from:** heating (water / central heating) **[1]**; produce electricity **[1]**; make sound **[1]**; make light **[1]**
3. The minimum energy required to start a reaction **[1]**
4. A correctly drawn reaction profile for an exothermic reaction **[1]**

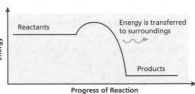

5. A correctly drawn reaction profile for an endothermic reaction **[1]**

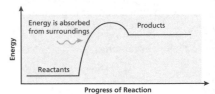

6. making chemical bonds **[1]**
7. Bond breaking: 432 + 155 = 587kJ/mol **[1]**; bond making: 2 × 565 = 1130kJ/mol **[1]**; bond breaking – bond making (ΔH) = 587 – 1130 = –543kJ/mol **[1]**; the reaction is exothermic **[1]**
8. a) Endothermic **[1]**
   b) Exothermic **[1]**
   c) Endothermic **[1]**
   d) Endothermic **[1]**
9. C **[1]**

Page 44 Types of Chemical Reactions

1. A **[1]**; C **[1]**

   > Remember, oxidisation is the addition of oxygen.

2. Oxidation is loss of electrons **[1]**; and reduction is gain of electrons **[1]**
3. a) $2Na(s) + Cl_2(g) \rightarrow 2NaCl(s)$ **[2]** (1 mark for correct balancing; 1 mark for correct state symbols); sodium is oxidised and chlorine is reduced **[1]**
   b) $2Mg(s) + O_2(g) \rightarrow 2MgO(s)$ **[2]** (1 mark for correct balancing; 1 mark for correct state symbols); magnesium is oxidised and oxygen is reduced **[1]**
   c) $2Li(s) + Br_2(g) \rightarrow 2LiBr(s)$ **[2]** (1 mark for correct balancing; 1 mark for correct state symbols); lithium is oxidised and bromine is reduced **[1]**

d) $CuO(s) + H_2(g) \rightarrow Cu(s) + H_2O(l)$ **[2]** (1 mark for correct balancing; 1 mark for correct state symbols); copper is reduced and hydrogen is oxidised **[1]**

4. a) $H^+(aq)$ **[1]**
   b) $OH^-(aq)$ **[1]**
5. acid + base $\rightarrow$ salt + water **[1]**
6. a) $H_2SO_4(aq) + 2NaOH(aq) \rightarrow$ $Na_2SO_4(aq) + 2H_2O(l)$ **[2]** (1 mark for correct reactants; 1 mark for correct products)
   b) Na^+ **[1]**; SO_4^{2-} **[1]**
   c) $H^+(aq) + OH^-(aq) \rightarrow H_2O(l)$ **[2]** (1 mark for correct ions; 1 mark for correct product)

Page 45 pH, Acids and Neutralisation
1. An acid that does not fully dissociate when dissolved in water **[1]**
2. a) $2mol/dm^3$ H_2SO_4 **[1]**
   b) $3mol/dm^3$ HNO_3 **[1]**
3. 1000 times greater (10 × 10 × 10 or 10^3) **[1]**

Page 45 Electrolysis
1. a) Cations **[1]**
   b) Anions **[1]**
2. The process of breaking down ionic compounds into simpler substances using an electric current **[1]**
3. Table salt is a solid at room temperature and pressure and electrolysis only works if the ion is in solution or molten **[1]**
4. Set up an electrolytic cell using a nail as the cathode **[1]**; and copper for the anode **[1]**; fill with copper(II) sulfate solution and apply an electric current **[1]**
5. a) Because they do not react with the products of electrolysis or the electrolyte **[1]**
   b) Platinum electrodes are very expensive / the same results can be achieved using cheaper electrodes **[1]**

Page 47 Quick Test
1. Iodine
2. The outermost electron in caesium is much further away from the nucleus than in lithium, so the force of attraction is weaker (it can lose this electron more easily)
3. Lithium

Page 49 Quick Test
1. Insert a lit splint into a test tube of the gas. If it makes a squeaky pop, it is hydrogen.
2. Warm with dilute sodium hydroxide solution. If a pungent gas is given off that turns damp red litmus paper blue, ammonium ions are present.
3. iron(II) ions / Fe^{2+}

Page 51 Quick Test
1. K^+
2. Crimson

3. **Any two from:** more sensitive; more accurate; able to measure factors more rapidly; can be used on very small samples

Page 53 Quick Test
1. titre = final volume – start volume
2. $concentration = \dfrac{moles}{volume}$
3. The volume occupied by 1 mole of a gas at room temperature and pressure, i.e. $24dm^3$.

Page 55 Quick Test
1. a) 320 tonnes / 320 000kg
   b) 89.9%

Page 57 Quick Test
1. Rate of reaction increases
2. Increase the pressure
3. A large block of calcium carbonate has a low surface area to volume ratio compared to powder, which consists of lots of small particles with a high surface area to volume ratio.

Page 59 Quick Test
1. The steeper the line / the greater the gradient, the faster the rate of reaction
2. Catalysts reduce the activation energy (making a reaction more likely)
3. Enzymes

Page 61 Quick Test
1. Temperature; pressure; concentration
2. False
3. There will be a decrease in the yield of products and an increase in amount of reactants, restoring the original conditions according to Le Chatelier's principle.

Page 62 Introducing Chemical Reactions
1. a) H = 2, S = 1, O = 4 **[1]**
   b) Cu = 1, N = 2, O = 6 **[1]**
   c) C = 3, H = 6, O = 2 **[1]**
   d) C = 2, H = 6 **[1]**
2. a) $CuO(s) + H_2SO_4(aq) \rightarrow$ $CuSO_4(aq) + H_2O(l)$ **[2]** (1 mark for correct reactants; 1 mark for correct products)
   b) $2Mg(s) + O_2(g) \rightarrow 2MgO(s)$ **[2]** (1 mark for correct reactants; 1 mark for correct products)
   c) $Mg(OH)_2(aq) + 2HCl(aq) \rightarrow$ $MgCl_2(aq) + 2H_2O(l)$ **[2]** (1 mark for correct reactants; 1 mark for correct products)
   d) $CH_4(g) + 2O_2(g) \rightarrow CO_2(g) + 2H_2O(l)$ **[2]** (1 mark for correct reactants; 1 mark for correct products)
3.

Name of Ion	Formula
Carbonate	CO_3^{2-} **[1]**
Lithium	Li^+ **[1]**
Iron(III) **[1]**	Fe^{3+}
Oxide **[1]**	O^{2-}
Sulfate	SO_4^{2-} **[1]**

4. a) $Pb(s) \rightarrow Pb^{2+}(aq/l) + 2e^-$ **[1]**
   b) $Al^{3+}(aq/l) + 3e^- \rightarrow Al(s)$ **[1]**
   c) $Br_2(l) + 2e^- \rightarrow 2Br^-(aq)$ **[1]**
   d) $Ag^+(aq) + e^- \rightarrow Ag(s)$ **[1]**

Page 63 Chemical Equations
1. The reacting species **[1]**

> Spectator ions are not included in ionic equations.

2. a) $Ag^+(aq) + Cl^-(aq) \rightarrow AgCl(s)$ **[1]**
   b) $Mg^{2+}(aq) + CO_3^{2-}(aq) \rightarrow MgCO_3(s)$ **[1]**

Page 63 Moles and Mass
1. a) $\dfrac{6.9g}{6.9g/mol} = 1mol$ **[1]**
   b) $\dfrac{62g}{31g/mol} = 2mol$ **[1]**
2. $(1 \times 14) + (4 \times 1) + (1 \times 35.5) = 53.5g/mol$ **[1]**
3. a) $\dfrac{183.8}{6.022 \times 10^{23}} = 3.05 \times 10^{-22}g$ **[1]** (Accept $3.01 \times 10^{-22}g$)
   b) $\dfrac{118.7}{6.022 \times 10^{23}} = 1.97 \times 10^{-22}g$ **[1]** (Accept $2.00 \times 10^{-22}g$)
4. $BaCl_2 + MgSO_4 \rightarrow BaSO_4 + MgCl_2$, 5mol of $BaCl_2$ makes 5mol of $BaSO_4$, 5mol of $BaSO_4 =$ $5 \times (137.3 + 32 + (4 \times 16))g/mol$ **[1]**; = $1166.5g$ **[1]**

> The stoichiometry of the reaction is 1 : 1 ratio reactant to product.

5. B **[1]**

Page 64 Energetics
1. a) Energy is taken in from the environment / surroundings **[1]**
   b) The energy comes from Atu's leg (heat energy) **[1]**
2. a) bond breaking: 436 + 243 = 679kJ/mol, bond making: 2 × 431 = 862kJ/mol **[1]**; bond breaking – bond making (ΔH) = 679 – 862 = –183kJ/mol **[1]**; the reaction is exothermic **[1]**
   b) A correctly drawn reaction profile for an exothermic reaction **[1]**

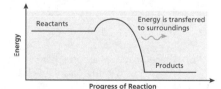

Page 65 Types of Chemical Reactions
1. A **[1]**; D **[1]**
2. a) Iron(III) oxide **[1]**
   b) Carbon monoxide **[1]**
   c) $Fe_2O_3(s) + 3CO(g) \rightarrow 2Fe(s) + 3CO_2(g)$ **[2]** (1 mark for correct reactants; 1 mark for correct products)
3. a) $2AgNO_3(aq) + Cu(s) \rightarrow$ $Cu(NO_3)_2(aq) + 2Ag(s)$ **[2]** (1 mark for correct reactants; 1 mark for correct products)

Answers

b) Silver [1]

c) Copper nitrate [1]

d) Silver is reduced as it gains an electron to become solid silver [1]; copper is oxidised as it loses two electrons to become an ion [1]

Page 66 pH, Acids and Neutralisation

1. a) $Mg(OH)_2(aq) + \underline{2HCl(aq)} \rightarrow$
   $MgCl_2(aq) + 2H_2O(l)$ [1]

   b) $\underline{H_2SO_4(aq)} + 2NaOH(aq) \rightarrow$
   $Na_2SO_4(aq) + 2H_2O(l)$ [1]

   c) $\underline{2CH_3COOH(aq)} + 2Na(s) \rightarrow$
   $H_2(g) + 2CH_3COONa(aq)$ [1]

   d) $\underline{2HF(aq)} + Mg(s) \rightarrow MgF_2(aq) + H_2(g)$ [1]

2. a) calcium + sulfuric acid → calcium sulfate + hydrogen [1]

   b) Above calcium [1]

3. a) $HNO_3(aq) + NaOH(aq) \rightarrow$
   $NaNO_3(aq) + H_2O(l)$ [2]
   (1 mark for correct reactants; 1 mark for correct products)

   b) $Na^+(aq)$ [1]; $NO_3^-(aq)$ [1]

   c) $H^+(aq) + OH^-(aq) \rightarrow H_2O(l)$ [2] (1 mark for correct ions; 1 mark for correct product)

4. A strong acid dissociates completely [1]

5. 10 000 ($10 \times 10 \times 10 \times 10$ or 10^4) [1]

Page 67 Electrolysis

1. Molten or in solution [1]

2. a) C [1]

   b) anion = oxygen [1]; cation = hydrogen [1]

   c) i) $2O^{2-}(aq) \rightarrow O_2(g) + 4e^-$ [1]
   ii) $2H^+(aq) + 2e^- \rightarrow H_2(g)$ [1]

3. $X = Cu^{2+}(aq)$ [1]; $Y = SO_4^{2-}$ (and OH^-) [1]

Pages 68–75 Practice Questions

Page 68 Predicting Chemical Reactions

1. C [1]

2. 1 [1]

3. fluorine, chlorine, bromine, iodine [1]

4. a) A correctly drawn sodium atom (2.8.1) [1]; and potassium atom (2.8.8.1) [1]

   b) Potassium moves around more violently [1]; and burns with a lilac flame [1]

   c) D [1]

5. Beryllium, Be [1]

Group 2 elements are more reactive going down the group.

6. $Zn^{2+}(aq)$, $H^+(aq)$, $SO_4^{2-}(aq)$ [1]

Page 69 Identifying the Products of Chemical Reactions

1. C [1]

2. Collect the gas in a test tube and bring a lit splint to the edge of the tube [1]; if it burns with a squeaky pop, hydrogen is present [1]

3. a) iron(II) / $Fe^{2+}(aq)$ [1]

   b) $FeSO_4(aq)$ [1]

4. $Cu^{2+}(aq)$ [1]; $OH^-(aq)$ [1]

5. $BaSO_4(s)$ [1]

Page 70 Ion Tests and Instrumental Methods of Analysis

1. C [1]

2. The pH stays more or less the same and then suddenly increases to pH 12 / 14 [1]

Acidic and alkali solutions do not change from acid to alkali or vice versa gradually. The solutions tend to be able to take a set amount of acid or alkali before changing abruptly from one to the other.

3. a) mass = number of moles × molar mass [1]; = 0.5mol × 120g/mol [1]; = 60g [1]

   b) The concentration would double [1]

4. a) B [1]

   b) A mixture of ions would give out different colours, which would make identifying the potassium ion (which produces a lilac flame) very difficult [1]

5. **Any two from:** more sensitive [1]; can detect and report far lower quantities [1]; more accurate [1]; able to measure factors more rapidly [1]; quantitative rather than qualitative [1]

Page 71 Monitoring Chemical Reactions

1. a) So that he can see the change from alkali to acid [1]

   b) Safety goggles must be worn and care should be taken when handling glassware [1]

   c) The indicator will change colour permanently [1]

2. B [1]

3. Amount of HCl (mol) = concentration × volume = 0.05 × 0.2 = 0.01mol [1]; the ratio of NaOH to HCl is 1:1, so number of moles of NaOH must also be 0.01mol [1]; concentration NaOH = $\frac{0.01}{0.035}$ = 0.29mol/dm³ (to 2 significant figures) [1]

4. a) $2.00 \times 24dm^3 = 48dm^3$ [1]

Remember, the molar volume of any gas at room temperature and pressure is 24dm³.

   b) The volume would increase, if the vessel the gas was in could expand [1]; if it could not, the pressure would increase [1]

5. B [1]

Page 72 Calculating Yields and Atom Economy

1. a) M_r of $MgSO_4$ = 120.4g/mol, $\frac{120g/mol}{1mol}$
   = 120.4g [1]

1 mole of MgO has a M_r of (24.3 + 16 =) 40.3g/mol. 40.3g of MgO is used, so 1mol of MgO used. This means that 1mol of $MgSO_4$ is made.

   b) $\frac{100g}{120.4g} \times 100 = 83.1\%$ [1]

2. A [1]

3. **Any three from:** atom economy, this will determine how efficient the reaction is at producing the desired product [1]; the environmental impact of the reaction, a pathway that doesn't produce any hazardous waste is likely to be chosen over one that does [1]; cost, if an efficient reaction is too expensive, as less costly but possibly less efficient route may be chosen [1]; rate of reaction, a slow reaction may be dismissed in favour of a faster pathway / if a reaction is too fast, this might be a problem, i.e. if it cannot be contained then there could be an explosion [1] Yield/ Equilibrium position – Ideally a reaction with a high yield or equilibrium position that favours the products will be chosen. (Only award a mark if an explanation is given for the factor)

Page 73 Controlling Chemical Reactions

1. a) C [1]

   b) The higher the temperature, the faster the rate of reaction [1]

2. B [1]; C [1]; D [1]

Page 74 Catalysts and Activation Energy

1. Catalysts lower the activation energy for a reaction [1]; this means that the reaction will happen more quickly when the catalyst is present [1]

2. a) Catalase is the catalyst, so it is written over the arrow and not as a reactant [1]

Remember, a catalyst is not used up in a reaction – it is not a reactant.

   b) Catalase is made of amino acids / is a protein / is an organic molecule [1]; manganese(IV) oxide is an inorganic molecule / catalyst [1]

3. a) i) A correct diagram showing particles at a low temperature [1]
   ii) A correct diagram showing particles at a high temperature [1]

Low Temperature

High Temperature

b) The faster reactant particles move, the greater the chance of successful collisions with other reactant particles [1]; successful collisions lead to product particles being formed [1]; also, there will be more reactant particles with energy greater than the activation energy [1]

Page 74 Equilibria

1. C [1]; D [1]
2. Nothing can enter or leave a closed system [1]; and the temperature and pressure remain the same [1]
3. C [1]
4. Temperature [1]; concentration [1]; pressure [1]
5. Le Chatelier's principle states that when the conditions of a system are altered, the position of the equilibrium changes [1]; to try to restore the original conditions [1]
6. a) The amount of product would increase [1]
   b) Nothing [1]
   c) More reactant ($CH_3COOH(aq)$) will be produced [1]
7. Lower temperature (as reaction is exothermic) [1]; increase concentration of reactants (as the product will be formed to counter that change) [1]

Pages 76–93 Revise Questions

Page 77 Quick Test

1. Heat (in a blast furnace) with carbon
2. The oxygen that forms at the anode reacts with the carbon producing $CO_2(g)$, which then escapes, so the anode gradually wears away.
3. **Any two from:** can be bred to tolerate the metal ion; they reproduce quickly; they are very efficient / only target a specific ion

Page 79 Quick Test

1. Hydrogen; nitrogen
2. The reaction would become too expensive (as it would require a lot of energy and all the necessary safety measures would add to the cost)
3. **Any three from:** the amount / price of energy used; labour costs; how quickly the new substance can be made; the cost of raw materials; cost of safely disposing of by-products

Page 81 Quick Test

1. Recycling is when the materials in an object are used for a new purpose – the refilled bottles are being reused, not recycled.
2. With some plastics, the materials may be contaminated, so it would be impractical / too expensive to recycle them.
3. In steel there is an element that exists between the metal atoms (it is an

interstitial alloy); in brass the atom of one element takes the place of an atom of the other element (it is a substitutional alloy).

Page 83 Quick Test

1. For sacrificial protection (the zinc reacts first with water and oxygen to protect the steel beneath it from corrosion)
2. To provide a barrier against water and oxygen and provide sacrificial protection.
3. Poly(styrene) is an insulator, so it keeps drinks hot / cold for longer (and is relatively cheap and light).

Page 85 Quick Test

1. C_3H_6
2. $C_nH_{2n+1}OH$
3. –COOH

Page 87 Quick Test

1. An alcohol
2. The monomers form a polymer and a small molecule, such as water or ammonia, is produced as a by-product.
3. They are made out of amino acid monomers

Page 89 Quick Test

1. The breaking down of long-chain hydrocarbons into smaller molecules
2. The longer the chain, the greater the total amount of intermolecular forces that have to be overcome
3. It is difficult to store hydrogen safely; it is difficult to move hydrogen around; the cells are expensive

Page 91 Quick Test

1. Volcanic activity
2. Photosynthesis
3. Methane

Page 93 Quick Test

1. The burning / incomplete combustion of fossil fuels
2. Chlorine kills any bacteria in the water
3. The contact process; the Haber process

Pages 94–101 Review Questions

Page 94 Predicting Chemical Reactions

1. C [1]
2. B [1]
3. caesium, rubidium, potassium, sodium, lithium [1]
4. a) i) 2.1 [1]
      ii) 2.8.2 [1]
      iii) 2.8.3 [1]
      iv) 2.6 [1]
   b) More bubbles should be drawn than in the Mg tube (at least 5 more) to show a more vigorous reaction [1]
   c) Hydrogen [1]

Page 95 Identifying the Products of Chemical Reactions

1. $NH_4^+(aq)$ [1]; $OH^-(aq)$ [1]
2. A [1]

3. Collect the gas in a test tube and place a glowing splint into the tube [1]; if the splint relights, the gas is oxygen [1]
4. To test for iron(II) react with sodium hydroxide [1]; if a green precipitate is formed, iron(II) ions are present [1]; to test for sulfate ions, add hydrochloric acid then mix with barium chloride [1]; if a white precipitate forms, add more hydrochloric acid [1]; if the precipitate remains, the anion was sulfate [1]
5. a) $Ba^{2+}(aq) + CO_3^{2-}(aq) \rightarrow BaCO_3(s)$ [2]
      (1 mark for the correct ions; 1 mark for the correct product)
   b) $Ca^{2+}(aq) + 2OH^-(aq) \rightarrow Ca(OH)_2(s)$ [2]
      (1 mark for the correct ions; 1 mark for the correct product)
6. Collect the gas in a tube and place damp indicator paper into the top of the tube [1]; if the indicator paper is bleached (turns white), the gas is chlorine [1]

Page 96 Ion Tests and Instrumental Methods of Analysis

1. Carbon dioxide – turns limewater milky white [1]; Hydrogen – burns with a squeaky pop [1]; Oxygen – relights a glowing splint [1]
2. a) Dip a clean nichrome wire into concentrated hydrochloric acid [1]; dip the wire into the solid that is being tested [1]; place the wire with the sample into the hottest part of the Bunsen burner flame [1]; note the colour that is produced [1]
   b) Lithium / $Li^+(aq)$ [1]
3. a) Add a small amount of nitric acid to the solution (to acidify it) [1]; add silver nitrate [1]; if a halogen ion is present, an insoluble precipitate will be formed [1]
   b) Silver bromide, a cream coloured precipitate [1]
   c) $Ag^+(aq) + Br^-(aq) \rightarrow AgBr(s)$ [1]
4. She could use a data logger with a pH probe [1]

Page 97 Monitoring Chemical Reactions

1. a) Titre 1 = 18.80 [1]; Titre 2 = 17.90 [1]; Titre 3 = 18.00 [1]
   b) The result is anomalous (as this was the first titre used to find out the approximate position of the neutralisation) [1]
   c) $\dfrac{17.90 + 18.00}{2} = 17.95cm^3$ [2] (Award only 1 mark if the titre for Titration 1 is included in the calculation, giving an answer of $18.23cm^3$)

   Do not include any anomalous result when calculating a mean.

Answers

d) LiOH(aq) + HCl(aq) →
LiCl(aq) + H_2O(l) [1];
amount of HCl (mol) = concentration
× volume = 0.01795 × 0.2 =
0.00359mol [1]; the ratio of LiOH to
HCl is 1 : 1, so number of moles of
LiOH = 0.00359mol [1]; concentration
LiOH = $\frac{0.00359}{0.030}$ = 0.12mol/dm^3 (to 2
significant figures) [1]

2. a) X should be correctly marked on the
graph   [1]

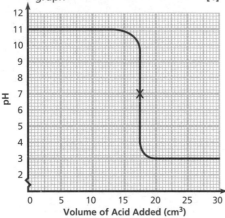

b) The end point is the point at which
the solution changes from alkaline
to acidic   [1]

Page 98 Calculating Yields and Atom Economy

1. Amount of HNO_3(aq) = concentration
× volume = 0.025 × 0.2 = 0.005mol [1];
the ratio of NaOH to HNO_3 is 1 : 1, so
number of moles of NaOH = 0.005mol
[1]; concentration NaOH = $\frac{0.005}{0.050}$ =
0.1mol/dm^3 (to 2 significant figures) [1]

2. a) Volume of gas = molar volume ×
number of moles of gas = 24dm^3 ×
3.00mol = 72dm^3 [1]

b) The volume would reduce (if the
container is able to change shape)
or the pressure would decrease   [1]

3. A   [1]

4. a) M_r: $(NH_2)_2$CO = (2 × 14) + (4 × 1) +
(1 × 12) + (1 × 16) = 60g/mol, mass
produced = 60g/mol × 1000mol =
60 000g / 60kg   [1]

> M_r of AgNCO = 108 + 14 + 12 + 16
> = 150g/mol.
> Number of moles used =
> $\frac{150\,000g}{150g/mol}$ = 1000mol.
> Therefore, reaction leads to
> 1000mol of $(NH_2)_2$CO.

b) 0.8 × 60kg = 48kg   [1]

Page 99 Controlling Chemical Reactions

1. a) B [1]; C [1]

b) A graph drawn with a line steeper
than the 25°C line   [1]

Volume of CO_2 Produced

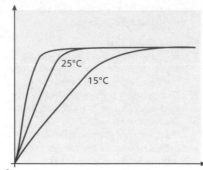

0   Time Taken for Cross to Disappear

Page 100 Catalysts and Activation Energy

1. a) A chemical that reduces the activation
energy, making it more likely that the
reaction will take place   [1]

b) Gases burned in a car engine are
harmful for the environment [1];
the catalytic convertor enables the
reactions to take place leading to
non-harmful gases and acceptable
levels of gases [1]

c) One graph line drawn showing the
reaction without a catalyst [1]; a second
correctly drawn graph line showing the
reaction with a catalyst [1]

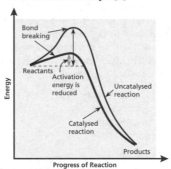

2. C   [1]

3. A [1]; C [1]

Page 100 Equilibria

1. forward [1]; HI [1]; reverse [1]; reverse [1]

2. B   [1]

3. B [1]; C [1]

4. More products   [1]

> ### Pages 102–109 Practice Questions

Page 102 Improving Processes and Products

1. B   [1]

2. a) copper(II) oxide + carbon →
copper + carbon dioxide [1]

b) For safety reasons / to prevent the
hot contents from being ejected
from the tube   [1]

3. D   [1]

4. B   [1]

5. Cu^{2+} + $2e^-$ → Cu(s)   [1]

6. a) Al^{3+}(l) and O^{2-}(l)   [1]

b) i) Al^{3+}(l) + $3e^-$ → Al(l)   [1]

ii) $2O^{2-}$(l) → O_2(g) + $4e^-$ [1]

---

7. **Any three from:** bacteria reproduce
rapidly [1]; the bacteria accumulate
a specific ion [1]; the process is very
efficient [1]; the process is much cheaper
than extracting metals in a furnace
(displacement) [1]

Page 103 The Haber Process

1. a) Nitrogen is inert / unreactive [1];
and most plants cannot use it in its
atmospheric / molecular form [1]

b) nitrogen + hydrogen ⇌ ammonia [1]

2. Although these conditions lead to the
highest percentage of ammonia in the
mixture, the reaction occurs at a slow
rate [1]; and the cost of maintaining
a high pressure for a long time is very
great [1]

3. **Any three from:** the pressure – the
higher the pressure, the higher the
plant cost [1]; the temperature – the
higher the temperature, the higher
the energy cost [1]; the catalysts –
catalysts can be expensive to buy,
but production costs are reduced
because they increase the rate of
reaction [1]; the number of people
needed to operate machinery –
automation reduces the wages bill [1];
the amount of unreacted material that
can be recycled – recycling reduces
costs [1]

4. Lab equipment usually glassware /
large-scale industrial equipment often
metal [1]; chemists work on each
reaction / automation is common [1]

Page 104 Life Cycle Assessments, Recycling and Alloys

1. B, D, G, H, I   [1]

2. a) Recycled means to put the materials
to another use / purpose   [1]

b) Extracting new materials uses a
lot of energy, costs money and
contributes to environmental
pollution [1]; throwing away
products means the materials /
resources are no longer available [1];
recycling means that the materials
that would have been lost are now
used in a different way [1]; this often
involves less energy, lower costs
and less waste than obtaining new
materials [1]

c) The cost (in money and pollution)
is greater than the benefit gained
from recycling the parts   [1]

3. A [1]; E [1]

Page 105 Using Materials

1. a) D   [1]

b) Four correctly drawn lines [3]
(2 marks for two correct lines;
1 mark for 1 correct line)
iron bridge – paint, iron bench –
plastic coating, iron roof –
galvanising, steel bicycle chain – oil

Answers

2. Glass is inert / will not react with the chemicals [1]
3. a) Composite are mixtures or layers of materials chemically bonded together [1]
   b) **Any two from:** carbon fibre will not rust [1]; carbon fibre is much stronger [1]; carbon fibre is lighter [1]; carbon fibre is more durable [1]

Page 106 Organic Chemistry
1. a) B and F [1]
   b) The presence of a double bond between two carbon atoms (C=C) [1]
   c) A and C [1]
   d) C [1]
   e) [1]

   $$H-\underset{\underset{H}{|}}{\overset{\overset{H}{|}}{C}}-\underset{\underset{H}{|}}{\overset{\overset{H}{|}}{C}}-H$$

   f) Ethanol [1]
2. The longer-chain hydrocarbons contain a higher number of carbon atoms [1]; they will require more oxygen to burn [1]; so incomplete combustion often occurs and soot is formed [1]
3. a) alcohol + potassium manganate(VII) → carboxylic acid + manganese(IV) oxide + potassium hydroxide + water [1]
   b) The colour will change from purple to colourless [1]

Page 107 Organic Compounds and Polymers
1. a) Nucleotide [1]
   b) A base, sugar and phosphate [1]
   c) A correctly drawn nucleotide [1]

   P = Phosphate
   S = Sugar
   B = Base
2. C [1]

Page 108 Crude Oil and Fuel Cells
1. a) W = paraffin, X = diesel, Y = crude oil, Z = bitumen [3]
   (2 marks for two correct; 1 mark for one correct)
   b) At the top of the column [1]
2. Paraffin [1]; fuel oil [1]

Page 108 Interpreting and Interacting with Earth's Systems
1. a) As carbon dioxide levels increase, so does air temperature [1]
   b) Carbon dioxide is a greenhouse gas [1]; it traps heat energy in the atmosphere, leading to an increase in air temperature [1]
   c) A [1]; B [1]; D [1]
2. a) 20.96% [1] (Accept 21%)
   b) Photosynthesis leads to the reduction of carbon dioxide [1]; and the production of oxygen [1]

---

   c) Ammonia [1]; carbon dioxide [1]

Page 109 Air Pollution, Potable Water and Fertilisers
1. a) Carbon monoxide binds to the red blood cells instead of oxygen [1]; starving body cells of oxygen [1]
   b) Burning / Incomplete combustion of fossil fuels [1]
2. Oxides of nitrogen can form acid rain [1]; and photochemical smog [1]
3. C [1]

Pages 110–117 Review Questions

Page 110 Improving Processes and Products
1. A [1]
2. a) The carbon / graphite anode [1]
   b) C [1]

Page 111 The Haber Process
1. C [1]
2. B [1]; D [1]
3. a) A temperature of 450°C [1]; pressure of 200 atmospheres (atm) [1]; and an iron catalyst [1]
   b) A higher temperature (the optimum) is quicker but produces a lower yield [1]; a lower pressure than the optimum is cheaper [1]
4. **Any three from:** the price of energy (gas and electricity) [1]; labour costs (wages for employees) [1]; how quickly the new substance can be made (cost of catalyst) [1]; the cost of starting materials (reactants) [1]; the cost of equipment needed (plant and machinery) [1]

Page 112 Life Cycle Assessments, Recycling and Alloys
1. a) B, D, C, A [1]
   b) Less waste [1]; less fresh water used in its manufacture [1]
   c) **Any two from:** more energy is required to melt the ore (bauxite / aluminium oxide) [1]; anode reacts with oxygen to form carbon dioxide [1]; more fossil fuel used in production [1]
   d) Cork [1]
   e) Aluminium is easily recycled [1]

Page 113 Using Materials
1. **Any two from:** coating with plastic [1]; painting [1]; oiling [1]; galvanising [1]; sacrificial protection [1]
2. A composite material is made of a mixture of materials chemically bonded together [1]; the Roman concrete is made from stones bound together with cement [1]

Page 113 Organic Chemistry
1. a) A and F [1]; there are no double bonds / only single bonds between C atoms [1]
   b) A, B, C, E, F [1]
   c) B, C and E [1]
   d) Carboxylic acid [1]

---

   e) COOH and C=C [1]
2. B [1]

Page 114 Organic Compounds and Polymers
1. a) [1]

   $$\underset{\underset{H}{\displaystyle|}}{\overset{\overset{H}{\displaystyle|}}{C}}=\underset{\underset{H}{\displaystyle|}}{\overset{\overset{CH_3}{\displaystyle|}}{C}}$$

   b) There is a side CH_3 / methyl chain in poly(propene) [1]
2. a) Propanol [1]
   b) i) Propanoic acid [1]
      ii) It decolorises from purple [1]
3. [1]

   $$H-\underset{\underset{Br}{|}}{\overset{\overset{H}{|}}{C}}-\underset{\underset{Br}{|}}{\overset{\overset{H}{|}}{C}}-H$$

Page 115 Crude Oil and Fuel Cells
1. a) [1]

   $$\underset{\underset{F}{|}}{\overset{\overset{F}{|}}{C}}=\underset{\underset{F}{|}}{\overset{\overset{F}{|}}{C}}$$

   b) $C_{14}H_{30}$ [1]
   c) $C_{16}H_{34} \rightarrow C_{14}H_{30} + C_2H_4$ [1]
2. petrol [1]

Page 115 Interpreting and Interacting with Earth's Systems
1. a) Carbon dioxide is a greenhouse gas [1]; increased carbon dioxide levels leads to greater levels of heat energy being trapped in the atmosphere [1]; this results in increased air temperature and climate change [1]
   b) Levels of CO_2 have risen faster than was expected [1]; so reducing the levels will require more wells [1]
   c) 60.4% increase [1]
   d) Carbon dioxide is released through the burning of fossil fuels in factories and cars [1]; large-scale deforestation has removed trees that would normally take in carbon dioxide for photosynthesis [1]
2. A [1]

Page 117 Air Pollution, Potable Water and Fertilisers
1. Carbon monoxide is similar in shape to O_2 [1]; it binds with haemoglobin [1]; preventing O_2 from being transported to the cells [1]
2. B [1]
3. Desalination [1]; reverse osmosis [1]
4. a) A partially permeable membrane / semi-permeable membrane [1]
   b) In places that have a large population [1]; but a low rainfall / poor water supply [1]
5. a) Nitrogen (N) / ammonia [1]; phosphoric acid / phosphorus (P) [1]; potassium (K) [1]
   b) Neutralising phosphoric acid with ammonia [1]

Answers

1. The time taken to react increases from A to D **[1]**; this means that the more concentrated the acid, the faster the reaction **[1]**; in terms of particles, there are more ions available per unit volume in A than in D **[1]**; so there are more likely to be successful collisions **[1]**

2. a) D **[1]**
   b) The line should not move – if it had been drawn in pen, the ink would have moved with the solvent front. **[1]**
   c) Ink 3 **[1]**
   d) 13/40 = 0.33 (Accept answers between 0.30 and 0.35) **[1]**

3. A **[1]**

4. a) Graphite has a giant covalent structure in which each carbon atom is bonded to three other carbon atoms **[1]**; the graphite structure is in layers, held together by weak forces, so the layers separate easily **[1]**; diamond has a giant covalent structure in which each carbon atom is covalently bonded to four other carbon atoms / the maximum number of carbon atoms **[1]**; this makes it very strong **[1]**
   b) Allotrope **[1]**
   c) Burn / react diamond and graphite with oxygen **[1]**; to show that they produce the same product: carbon dioxide **[1]**

5. a) Covalent bonds **[1]**; the dot and cross diagram shows that the electrons are shared / the ball and stick model and the displayed formula use a single lines or sticks to represent a covalent bond **[1]**
   b) The bond angles are incorrect **[1]**; the model does not show the 3D shape of a molecule **[1]**

6. a) Lithium is higher in the reactivity series than carbon, so it cannot be displaced by it – it can only be extracted by electrolysis **[1]**; iron is lower than carbon in the reactivity series, so it can be displaced by it **[1]**
   b) i) $Li^+(l) + e^- \rightarrow Li(l)$ **[2]** (1 mark for the correct ion; 1 mark for the correct product)
      ii) For electrolysis to work the ions have to be mobile (in solution / molten) / in a solid, the ions are not free to move **[1]**

7. 13.5cm³/min **[1]**

8. Containing one or more double bonds between carbon atoms (C=C) **[1]**

9. a) Particle A: 7 neutrons **[1]**; Particle B: atomic number of 20 **[1]**; Particle C: atomic mass of 16 **[1]**; and electronic structure 2.6 **[1]**; Particle D: 9 protons **[1]**
   b) D is an atom because the number of electrons equals the number of protons (the overall charge is neutral) **[1]**; B is an ion because it has two electrons less than the number of protons (so it has an overall charge) **[1]**
   c) It is in Group 5 **[1]**; and Period 2 **[1]**; the electronic configuration shows the numbers of electrons in each shell / energy level – the number of shells (two) gives the period number **[1]**; the number of electrons in the outermost shell / energy level (five) gives the group number **[1]**
   d) 20 **[1]**

10. C **[1]**

11. The number of neutrons varies **[1]**; as neutrons have mass this means that the atomic mass changes **[1]**; whilst the number of electrons stays the same **[1]**; as the number of protons is the same in each isotope **[1]**

12. Diesel **[1]**

13. a) 0.9g + 7.1g = 8.0g **[1]**
    b) 0.1mol of X, 0.1mol of chlorine / Cl_2 and 0.1mol of X chloride **[1]**

    > **X** has a relative atomic mass of 9g/mol and Devlin reacts 0.9g with chlorine. Start by working out the number of moles actually reacted.

    c) $X + Cl_2 \rightarrow XCl_2$ **[2]** (1 mark for the correct reactants; 1 mark for the correct product)

14. C **[1]**

15. a) $MgO(s) + H_2SO_4(aq) \rightarrow MgSO_4(aq) + H_2O(l)$ **[1]**
    b) **Any one from:** wear eye protection / safety goggles **[1]**; beware of spitting acid when heating the mixture **[1]**; do not boil the mixture **[1]**
    c) i) Keep adding magnesium oxide until no more dissolves (all acid has reacted) **[1]**
       ii) Pass the mixture through a filter funnel to remove unreacted magnesium oxide **[1]**

16. a) The nucleus **[1]**
    b) Protons and neutrons **[1]**
    c) B and C **[1]**
    d) B = boron **[1]**; C = carbon **[1]**

17. a) C_2H_4O **[1]**
    b) $(4 \times 12) + (8 \times 1) + (2 \times 16) = 88$ **[1]**

18. B **[1]**

19. a) The compound is ionic **[1]**; the ions of calcium and fluoride are arranged in a giant lattice structure in the solid form of the ion **[1]**; this means that it is extremely difficult to break the strong electrostatic forces between ions to melt it **[1]**
    b) Methane has a simple molecular structure **[1]**; it only has very weak intermolecular forces between molecules, so is easy to separate from other methane molecules **[1]**; so the melting point and boiling points are low **[1]**
    c) Correctly drawn lithium ion **[1]**; and oxide ion **[1]**

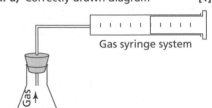

20. a) A substance that contains only one type of atom or molecule **[1]**
    b) In everyday life, pure means a substance comes from one source with no other contaminants **[1]**

21. a) Be **[1]**
    b) F **[1]**
    c) 29 **[1]**

22. C **[1]**

23. B **[1]**

24. a) Correctly drawn diagram **[1]**

Gas syringe system

    b) Add a catalyst **[1]**
    c) Wear eye protection **[1]**

25. a) E **[1]**; because (like B) it has a higher melting point than 100°C so will carry the oil **[1]**; but (unlike B) is insoluble in oil **[1]**
    b) Monomer **[1]**
    c) Correctly drawn unit **[1]**; three joined together in a chain **[1]**

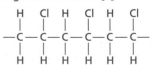

26. a) Calcium gives off a yellow-red colour and lithium a crimson colour – these are too similar to be able to tell them apart properly if they appear together **[1]**
    b) Hydrogen because the solution is acidic / pH 4 **[1]**; carbonate because the precipitate disappears with hydrochloric acid **[1]**; bromide as there is a cream coloured precipitate **[1]**

27. B **[1]**

28. C **[1]**

29. D **[1]**

30. a) $2NaHCO_3(s) \rightarrow Na_2CO_3(s) + H_2O(g) + CO_2(g)$ **[2]** (1 mark for the correct reactants; 1 mark for the correct products)

b) HCO_3^- (negative) [1]

c) Add barium chloride followed by dilute hydrochloric acid [1]; carbon dioxide is given off and any precipitate that initially forms will disappear [1]

31. a) The removal of salt from seawater to produce pure water [1]

b) Na_2SO_4 [1]

c) Chloride ions must be present in the water [1]

d) Fresh water is needed for people to survive [1]; if the water is contaminated with disease-causing organisms or high in salts then the person drinking it is likely to become ill and die [1]

32. **Any two from:** reactants are cheaper [1]; no waste products to deal with [1]; the yield is very high [1]

33. a) Nanoparticles have a very high surface area for reactants to adhere to [1]

b) The nanoparticles are small enough to enter the respiratory system and potentially cause health problems [1]

34. a) C = 2, H = 6, O = 1 [1]

b) [1]

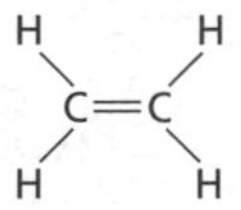

c) [1]

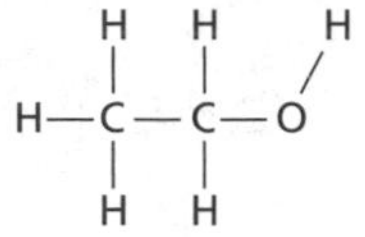

d) –OH [1]

35. metal + acid → salt + hydrogen [1]

36. a) magnesium nitrate + hydrogen [1]

b) zinc sulfate + hydrogen [1]

c) iron(II) chloride + hydrogen [1]

37. metal carbonate + acid → salt + carbon dioxide + water [1]

38. a) magnesium nitrate + carbon dioxide + water [1]

b) copper sulfate + carbon dioxide + water [1]

c) sodium chloride + carbon dioxide + water [1]

Periodic Table

Key
atomic number
symbol
name
relative atomic mass

(1)	(2)											(3)	(4)	(5)	(6)	(7)	(0)
																	2 **He** helium 4.0
1 **H** hydrogen 1.0																	
3 **Li** lithium 6.9	4 **Be** beryllium 9.0											5 **B** boron 10.8	6 **C** carbon 12.0	7 **N** nitrogen 14.0	8 **O** oxygen 16.0	9 **F** fluorine 19.0	10 **Ne** neon 20.2
11 **Na** sodium 23.0	12 **Mg** magnesium 24.3											13 **Al** aluminum 27.0	14 **Si** silicon 28.1	15 **P** phosphorus 31.0	16 **S** sulfur 32.1	17 **Cl** chlorine 35.5	18 **Ar** argon 39.9
19 **K** potassium 39.1	20 **Ca** calcium 40.1	21 **Sc** scandium 45.0	22 **Ti** titanium 47.9	23 **V** vanadium 50.9	24 **Cr** chromium 52.0	25 **Mn** manganese 54.9	26 **Fe** iron 55.8	27 **Co** cobalt 58.9	28 **Ni** nickel 58.7	29 **Cu** copper 63.5	30 **Zn** zinc 65.4	31 **Ga** gallium 69.7	32 **Ge** germanium 72.6	33 **As** arsenic 74.9	34 **Se** selenium 79.0	35 **Br** bromine 79.9	36 **Kr** krypton 83.8
37 **Rb** rubidium 85.5	38 **Sr** strontium 87.6	39 **Y** yttrium 88.9	40 **Zr** zirconium 91.2	41 **Nb** niobium 92.9	42 **Mo** molybdenum 95.9	43 **Tc** technetium	44 **Ru** ruthenium 101.1	45 **Rh** rhodium 102.9	46 **Pd** palladium 106.4	47 **Ag** silver 107.9	48 **Cd** cadmium 112.4	49 **In** indium 114.8	50 **Sn** tin 118.7	51 **Sb** antimony 121.8	52 **Te** tellurium 127.6	53 **I** iodine 126.9	54 **Xe** xenon 131.3
55 **Cs** cesium 132.9	56 **Ba** barium 137.3	57–71 lanthanides	72 **Hf** hafnium 178.5	73 **Ta** tantalum 180.9	74 **W** tungsten 183.8	75 **Re** rhenium 186.2	76 **Os** osmium 190.2	77 **Ir** iridium 192.2	78 **Pt** platinum 195.1	79 **Au** gold 197.0	80 **Hg** mercury 200.5	81 **Tl** thallium 204.4	82 **Pb** lead 207.2	83 **Bi** bismuth 209.0	84 **Po** polonium	85 **At** astatine	86 **Rn** radon
87 **Fr** francium	88 **Ra** radium	89–103 actinides	104 **Rf** rutherfordium	105 **Db** dubnium	106 **Sg** seaborgium	107 **Bh** bohrium	108 **Hs** hassium	109 **Mt** meitnerium	110 **Ds** darmstadtium	111 **Rg** roentgenium	112 **Cn** copernicium		114 **Fl** flerovium		116 **Lv** livermorium		

Formulae of Common Ions

It is useful to learn these, although the charges can be worked out if necessary.

Positive Ions (Cations)		Negative Ions (Anions)	
Name	Formula	Name	Formula
Hydrogen	H^+	Chloride	Cl^-
Sodium	Na^+	Bromide	Br^-
Silver	Ag^+	Fluoride	F^-
Potassium	K^+	Iodide	I^-
Lithium	Li^+	Hydroxide	OH^-
Ammonium	NH_4^+	Nitrate	NO_3^-
Barium	Ba^{2+}	Oxide	O^{2-}
Calcium	Ca^{2+}	Sulfide	S^{2-}
Copper(II)	Cu^{2+}	Sulfate	SO_4^{2-}
Magnesium	Mg^{2+}	Carbonate	CO_3^{2-}
Zinc	Zn^{2+}	Hydrogen carbonate	HCO_3^-
Lead	Pb^{2+}		
Iron(II)	Fe^{2+}		
Iron(III)	Fe^{3+}		
Aluminium	Al^{3+}		

Glossary and Index

Electron   a subatomic particle with a negative charge; occupy shells / energy levels around the nucleus   **46–47**

Electronic structure   a written indication of the numbers of electrons in each shell surrounding the nucleus of an atom   **12–13**

Emissions   the by-products of the combustion of fossil fuels   **88–89**

Endothermic   accompanied by the absorption of heat   **28–29, 60–61**

Enhanced greenhouse effect   the effect caused by the non-natural increase in the amount of greenhouse gases in the atmosphere, leading to greater temperatures than should occur naturally   **90–91**

Environment   the natural world, as a whole or in a particular geographical area   **28–29**

Enzymes   biological catalysts that can reduce activation energy   **58–59**

Equilibrium   a balance; occurs in a reversible reaction when the rate of the forward reaction is equal to the rate of the reverse reaction   **60–61**

HT Equilibrium mixture   a mixture where there is a balance between reactants forming new products and the products forming new reactants   **32–33**

Exothermic   accompanied by the production of heat   **28–29, 60–61**

F

Formulae   the way of recording the number of atoms chemically bonded together in reactants and products, using element symbols and numbers   **22–23**

Formulation   a mixture carefully designed to have specific properties   **10–11**

Fossil fuels   fuels produced from oil, coal and natural gas that took millions of years to form and so cannot be replaced   **90–91**

Fractional distillation   the splitting of crude oil into different fractions based on boiling points   **88–89**

Fuel cell   a device that converts the chemical energy from a fuel into electricity through a chemical reaction of positively charged hydrogen ions with oxygen   **88–89**

Fullerenes   a form of carbon generally having a large spheroidal molecule consisting of a hollow cage of 60 or more carbon atoms   **16–17**

Functional group   a specific group of atoms or bonds within a molecule that is responsible for the characteristic chemical reactions of that molecule   **84–85**

G

Gas   a state of matter where the atoms or molecules are free to move, so that they will take up the available space in any container   **22–23**

Giant covalent structure   a giant structure in which the atoms are all joined together by strong covalent bonds   **14–15**

Graphene   a form of carbon consisting of planar sheets one atom thick, with the atoms arranged in a honeycomb-shaped lattice   **16–17**

Greenhouse effect   the process by which certain gases cause infrared radiation to be trapped, causing the temperature of the atmosphere to be higher than if the infrared radiation could escape into space   **90–91**

Group   a column of elements in the periodic table, each element having the same number of electrons in the outermost shell orbiting the nucleus   **12–13**

H

Haber process   a method of fixing atmospheric nitrogen into ammonia   **78–79**

HT Half equation   an equation written to describe an oxidation or reduction half-reaction   **24–25**

Halide   a halogen ion with a single negative charge   **48–49**

Halogen   an element in Group 7 of the periodic table   **46–47**

Homologous series   a series of organic compounds with a similar general formula, possessing similar chemical properties due to the presence of the same functional group   **84–85**

Hydrocarbon   a molecule containing only carbon and hydrogen atoms   **84–85**

HT Hyperaccumulators   plants that can contain over 100 times higher metal concentrations than non-accumulator plants grown under the same conditions   **76–77**

I

Inert electrode   an electrode which will not be affected by electrolysis   **34–35**

Infrared radiation   electromagnetic radiation that heats objects   **90–91**

Inorganic   a molecule which does not contain carbon   **58–59**

Instrumental analysis   a quantitative method of measuring a variable by using a machine   **50–51**

Intermolecular force   the sum of all the forces between two neighbouring molecules   **16–17**

Interstitial   in the spaces between atoms   **80–81**

Ion   an element or molecule that has a net charge, having gained or lost an electron or electrons   **8–9, 24–25**

Ionic bond   a chemical bond between two ions where one ion donates electrons and the other receives electrons   **12–15**

Irreversible   cannot be made to go the other way; reactant → product   **60–61**

Isotopes   versions of an element that contain the same number of protons but have different numbers of neutrons in their nuclei, and hence have a different relative atomic mass   **8–9**

K

Kinetic energy   the movement energy of an object, e.g. for a particle   **56–57**

L

HT Le Chatelier's principle   a rule used to predict what will happen in a dynamic equilibrium if certain key conditions are changed   **60–61**

Life cycle assessment   a tool for working out the impact a product will have on the environment over its entire existence   **80–81**

Limiting reactant   the reactant that is completely used up in a reaction; it stops the reaction from going any further and any further products being produced.   **24–25**

Liquid   a state of matter where the atoms or molecules are packed together, leading to a stable volume but no definite shape   **22–23**

M

Mass number   the total number of protons and neutrons in the nucleus of an atom   **8–9**

Minerals   naturally occurring inorganic solids, with definite chemical compositions, and ordered atomic arrangements   **76–77**

Mobile phase   the liquid or gas that moves through the stationary phase in chromatography   **10–11**

Model   material, visual, mathematical or computer simulations of reality; used to help explain what is happening scientifically and make predictions   **14–15**

HT Molar volume   the volume of one mole of a substance   **52–53**

HT Mole   the amount of pure substance containing the same number of chemical units as there are atoms in exactly 12g of carbon-12   **26–27, 52–53**

Molecular formulae   formulae of a molecule showing the type and number of atoms   **84–85**

Molecule   a group of atoms chemically bonded together   **8–9**

Monomer   a subunit of a polymer   **86–87**

N

Nanoparticle   a particle that is in the size range 1–100 nm   **16–17**

HT Net ionic equation   an equation that shows only those species of chemical which change charge in a reaction   **24–25**

Neutralisation   the reaction between an acid and a base producing a neutral, pH 7 solution   **30–31, 32–33**

Collins

Chemistry

OCR GCSE Revision
Chemistry

OCR Gateway GCSE

Workbook

Eliot Attridge

Revision Tips

Rethink Revision

Have you ever taken part in a quiz and thought *'I know this!'* but, despite frantically racking your brain, you just couldn't come up with the answer?

It's very frustrating when this happens but, in a fun situation, it doesn't really matter. However, in your GCSE exams, it will be essential that you can recall the relevant information quickly when you need to.

Most students think that revision is about making sure you **know** stuff. Of course, this is important, but it is also about becoming confident that you can **retain** that *stuff* over time and **recall** it quickly when needed.

Revision That Really Works

Experts have discovered that there are two techniques that help with all of these things and consistently produce better results in exams compared to other revision techniques.

Applying these techniques to your GCSE revision will ensure you get better results in your exams and will have all the relevant knowledge at your fingertips when you start studying for further qualifications, like AS and A Levels, or begin work.

It really isn't rocket science either – you simply need to:

- **test yourself** on each topic as many times as possible
- **leave a gap** between the test sessions.

Three Essential Revision Tips

1. **Use Your Time Wisely**

   - Allow yourself plenty of time.
   - Try to start revising at least six months before your exams – it's more effective and less stressful.
   - Your revision time is precious so use it wisely – using the techniques described on this page will ensure you revise effectively and efficiently and get the best results.
   - Don't waste time re-reading the same information over and over again – it's time-consuming and not effective!

2. **Make a Plan**

   - Identify all the topics you need to revise (this All-in-One Revision & Practice book will help you).
   - Plan at least five sessions for each topic.
   - One hour should be ample time to test yourself on the key ideas for a topic.
   - Spread out the practice sessions for each topic – the optimum time to leave between each session is about one month but, if this isn't possible, just make the gaps as big as realistically possible.

3. **Test Yourself**

   - Methods for testing yourself include: quizzes, practice questions, flashcards, past papers, explaining a topic to someone else, etc.
   - This All-in-One Revision & Practice book provides seven practice opportunities per topic.
   - Don't worry if you get an answer wrong – provided you check what the correct answer is, you are more likely to get the same or similar questions right in future!

Visit our website to download your free flashcards, for more information about the benefits of these techniques, and for further guidance on how to plan ahead and make them work for you.

www.collins.co.uk/collinsGCSErevision

Contents

Particle Model and Atomic Structure

1   The smallest particle that retains the properties of the element is the:

A neutron   B proton   C atom   D Higgs boson

Your answer ☐ [1]

2   Which of the following correctly explains why an atom has no charge?

A The number of neutrons equals the number of protons.

B The number of protons equals the number of electrons.

C The number of neutrons equals the number of electrons.

D The total number of protons and neutrons equals the number of electrons.

Your answer ☐ [1]

3   This table about the three main isotopes of carbon, is missing some key information.

Isotope	Symbol	Mass Number	Atomic Number	Protons	Neutrons	Electrons
Carbon-12	$^{12}_{6}C$	12	6	B	6	D
A	$^{13}_{6}C$	13	6	6	7	6
Carbon-14	$^{14}_{6}C$	14	6	6	C	6

a) Fill in the missing sections of the table. [4]

b) What is the name of the particle whose number defines an element? [1]

c) Palladium has an atomic mass of 106 and atomic number of 46.

How many **neutrons** are in the nucleus? ☐ [1]

4   Look at the diagram of a gas. Particles are drawn as circles.

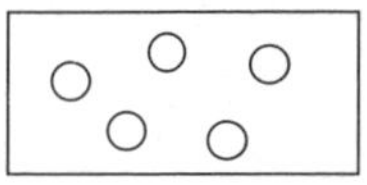

Explain **three** limitations of this model.

[3]

Total Marks _____ / 11

Purity and Separating Mixtures

1. Glucose, $C_6H_{12}O_6$, can be bought at the following price points:

   - Supermarket: £3/kg
   - Industrial chemical supplier: £6/kg; £18/kg; £114/kg

   Suggest why the chemical supplier can charge £114/kg for glucose.

   ....

   .... [2]

2. Which of the following is the empirical formula of glucose, $C_6H_{12}O_6$?

   A $C_6H_{12}O_6$ **B** $C_2H_3O_2$ **C** CHO   **D** CH_2O

   Your answer ☐ [1]

3. Calculate the relative formula mass of $Mg(OH)_2$. Show your working.

   relative formula mass = ... [2]

4. What is the empirical formula of a chemical containing 60g of carbon, 12g of hydrogen and 16g of oxygen? Show your working.

   empirical formula = ... [2]

5. Sandy wants to extract the pigment in ink so that she can transport it easily on a camping expedition.

   Explain how she can separate the pigment from the ink.

   ....

   ....

   .... [3]

   Total Marks .... / 10

Bonding

1   Tungsten is a metal. Which of the following is **not** a reason why it is a metal.

   A   It conducts electricity easily.

   B   It is malleable.

   C   It is a good conductor of heat.

   D   It has the symbol W.

   Your answer ☐ [1]

2   Which of the following shows the **electronic structure** for magnesium?

   A   2.8.2   B   2.8.1   C   2.8.8.2   D   2.8.8.1

   Your answer ☐ [1]

3   Explain the differences between a **covalent bond** and an **ionic bond**.

   ....

   ....

   ....

   ....

   [4]

4   Sodium, Na, is a metal. It has the atomic number 11.

   Which of the following can you **not** deduce using the atomic number?

   A   The electronic structure.   B   The period sodium is in.

   C   The number of isotopes.   D   The group sodium is in.

   Your answer ☐ [1]

5   Draw the electronic structure for the element chlorine, Cl.

   [1]

Total Marks _____ / 8

Models of Bonding

1 Methane has the formula CH_4.

a) Draw a dot and cross diagram and the displayed formula for methane in the spaces below.

Dot and Cross Diagram   **Displayed Formula**   [2]

b) Give **one** advantage that ball and stick models have over displayed formulae.

.... [1]

2 Aluminium forms an ion.

Draw a dot and cross diagram to show an aluminium cation.

[1]

3 Hydrogen sulfide, H_2S, is a covalent compound.

a) Draw a dot and cross diagram to show a molecule of H_2S.

[1]

b) How many electrons are there in a **single** covalent bond?

.... [1]

Total Marks .... / 6

Properties of Materials

1 Which of the following best describes an **allotrope**?

    A An element with the same number of protons and neutrons.

    B An element with the same number of protons but a different number of neutrons.

    C An element that can form different physical forms.

    D An element that can form long-chained molecules.

Your answer ☐ [1]

2 Graphite melts at 3600°C. Water melts at 0°C.

Explain why graphite and water melt at such widely different temperatures.

[4]

3 Kitchen chopping boards can be bought with embedded nanoparticles of silver.

a) What are nanoparticles?

[1]

b) Why are silver nanoparticles added to the chopping board?

[3]

4 Which of the following is **not** an allotrope of carbon?

    A glucose   **B** diamond   **C** buckyballs   **D** graphene

Your answer ☐ [1]

5 Why is diamond such a hard material?

[2]

Total Marks _____ / 12

Introducing Chemical Reactions

1 Write the number of atoms of each element for each formula below:

a) $BeAl_2O_4$

.... [1]

b) Al_2O_5Si

.... [1]

c) $Ba(PO_3)_2$

.... [1]

d) $CH_3(CH_2)_{16}COOH$

.... [1]

2 Copper oxide reacts with sulfuric acid.

Which of the following is the correct **word** equation for this reaction?

A copper oxide + sulfuric acid → copper sulfate

B copper oxide + sulfuric acid → copper sulfate + water

C copper oxide + sulfuric acid → copper + hydrogen sulfate

D copper oxide + sulfuric acid → copper sulfate + carbon dioxide + water

Your answer ☐ [1]

3 Magnesium carbonate reacts with nitric acid.

magnesium carbonate + nitric acid → magnesium nitrate + carbon dioxide + water

Write the **balanced symbol** equation for this reaction.

.... [2]

> **Total Marks** .... / 7

Chemical Equations

1 This question is about the four ions: **iron(II)**, **magnesium**, **sulfate** and **hydrogen carbonate**.

Which of the following correctly shows the charges on each ion?

	Iron(II)	Magnesium	Sulfate	Hydrogen carbonate
A	3+	2+	2+	2–
B	2+	3+	2–	1–
C	2+	2+	2–	1–
D	2+	2+	2–	1+

Your answer ☐ [1]

2 Give the definition of the term **stoichiometry**:

....

.... [2]

3 Write the **half equation** for the following:

a) Bromine ions to bromine gas.

.... [1]

b) Hydrogen gas to hydrogen ions.

.... [1]

c) Iron(III) ions to elemental iron.

.... [1]

4 Look at the following reaction:

sodium carbonate (aq) + copper sulfate (aq) ➜ copper carbonate (s) + sodium sulfate (aq)

a) What are spectator ions?

.... [1]

b) Which are the spectator ions in this reaction?

.... [1]

c) Write the net ionic equation for this reaction.

.... [2]

Total Marks _____ / 10

Moles and Mass

1. Which of the following is the best explanation of the chemical mole?

   A The amount of a chemical substance that is the same as in 12g of the element carbon.

   B The amount of a chemical substance that is the same as in 12g of the element carbon 12.

   C The amount of a chemical substance that is the same as in the element carbon.

   D The amount of a chemical substance that is the same as in 12g of the element hydrogen.

   Your answer ☐ [1]

2. Peter wants to measure out 3 moles of carbon atoms.

   Given that the A_r for carbon-12 is 12, which of the following shows the mass of carbon he should measure out?

   A 12g   B 2.4g   C 36g   D 3.6g

   Your answer ☐ [1]

3. Calculate the relative formula mass of the mineral chrysoberyl, $BeAl_2O_4$.
   Show your working.

   relative formula mass = _____ [3]

4. How many grams of water will be produced when 5 moles of hydrogen is completely combusted in air?
   Show your working.

   amount of water = _____ [3]

5. The A_r of rubidium is 85. What is the mass of a single Rb atom?

   mass = _____ [2]

Energetics

1 Hannah is going to a concert. On the way there, she passes a stall selling glow stick jewellery. She buys a glow stick bracelet.

Before the bracelet will glow, the two chemicals inside the bracelet have to come into contact with each other.

The chemicals, luminol and hydrogen peroxide, react and produce a chemical that emits light.

Explain why this reaction is classed as being **exothermic**.

....

....

....

[3]

2 Which of the following shows a reaction profile for an **endothermic** reaction?

      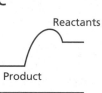

Your answer ☐ [1]

3 Describe the features of an **endothermic** reaction.

....

....

[2]

4 A series of reactions was carried out and the overall bond energies are recorded below.

For each, state whether the reaction is **exothermic** or **endothermic**.

a) −4kJ/mol [1]

b) +134kJ/mol [1]

c) −1kJ/mol [1]

d) +8kJ/mol [1]

Total Marks .... / 10

Types of Chemical Reactions

1 Draw a (circle) around the substance that is being **reduced** in each reaction.

   a) lithium + chlorine ➔ lithium chloride [1]

   b) beryllium + oxygen ➔ beryllium oxide [1]

   c) potassium + bromine ➔ potassium bromide [1]

   d) lead oxide + hydrogen ➔ lead + water [1]

2 Look at the following reaction:

   iron(III) oxide (aq) + carbon monoxide (g) ➔ iron (l) + carbon dioxide (g)

   a) Which species is being **reduced**?

   .... [1]

   b) Which species is being **oxidised**?

   .... [1]

   c) Write the **balanced symbol** equation for the reaction.

   .... [2]

3 Underline the **acid** in each of the following reactions.

   a) $MgCO_3(aq) + 2HCl(aq) ➔ MgCl_2(aq) + CO_2(g) + H_2O(l)$

   b) $HNO_3(aq) + NaOH(aq) ➔ NaNO_3(aq) + H_2O(l)$

   c) $2CH_3CH_2COOH(aq) + 2K(s) ➔ H_2(g) + 2CH_3CH_2COOK(aq)$

   d) $2HBr(aq) + Mg(s) ➔ MgBr_2(aq) + H_2(g)$ [4]

4 Which of the following represents a **low** pH?

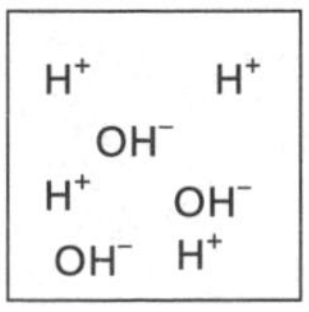

H^+ H^+
    OH^-
 H^+ OH^-
 OH^- H^+
	H^+ H^+
	 OH^- H^+ H^+
	 H^+ H^+
		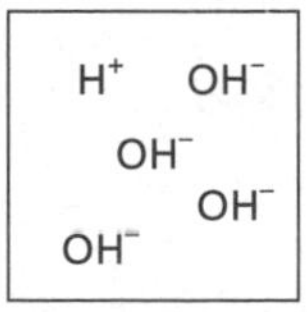 H^+ OH^-
		   OH^-
		      OH^-
		 OH^-
			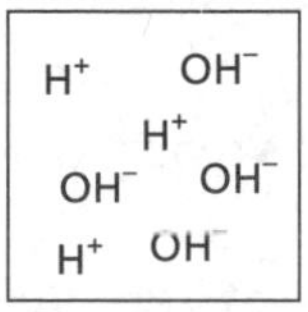 H^+ OH^-
			    H^+
			 OH^- OH^-
			 H^+ OH^-

A	**B**	**C**	**D**

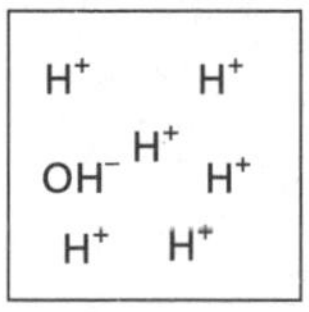

Your answer [ ] [1]

Total Marks .... / 13

pH, Acids and Neutralisation

1 Trevor is measuring the pH of a local river.
He uses universal indicator.

Explain why using universal indicator is **less** accurate than using a data logger with a pH probe.

[3]

2 Read the following statements about acids and alkalis.

Which statements are **true**? Write the letters below.

A Strong acids have a high pH.

B Strong acids dissociate easily.

C A dilute acid has more H^+ ions per unit volume than a concentrated acid.

D A dilute acid has fewer H^+ ions per unit volume than a concentrated acid.

E Weak acids form equilibria.

[3]

3 Look at the following representations of acids:

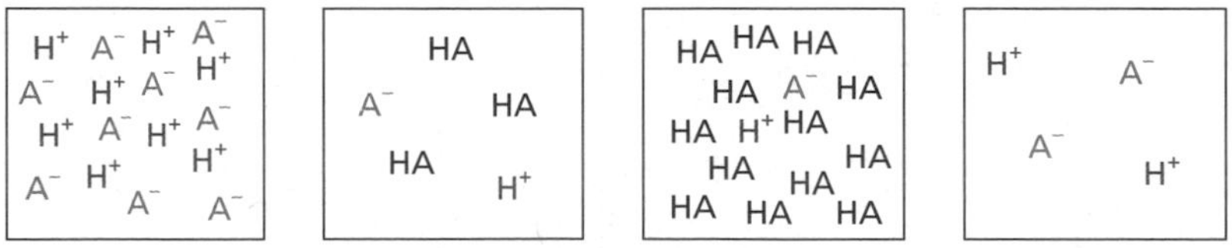

A   B   C   D

a) Which diagram represents a **concentrated, weak** acid? [1]

b) Which diagram represents a **dilute, strong** acid? [1]

c) Which diagram represents a **concentrated, strong** acid? [1]

d) Which diagram represents a **dilute, weak** acid? [1]

Total Marks _____ / 10

Electrolysis

1 When a metal is reacted with an acid it forms a metal salt, plus hydrogen gas.
A reactivity series is shown on the right.

Look at the following reactions:

W sodium + sulfuric acid ➔ sodium sulfate + hydrogen

X calcium + sulfuric acid ➔ calcium sulfate + hydrogen

Y potassium + sulfuric acid ➔ potassium sulfate + hydrogen

Z magnesium + sulfuric acid ➔ magnesium sulfate + hydrogen

Potassium
Sodium
Calcium
Magnesium
Aluminium
Carbon
Zinc
Iron
Tin
Lead
Hydrogen
Copper
Silver
Gold
Platinum

a) Which of the following shows the **correct order** of reactivity for these reactions, least to most reactive?

A WXYZ   **B** ZWXY   **C** ZXWY   **D** YWXZ

Your answer ☐ [1]

b) Element **V** can displace zinc from zinc oxide. It cannot be displaced by carbon.

Suggest where element **V** would appear in the reactivity series.

.... [1]

2 Explain why some metals have to be extracted from their ores by electrolysis.

....

....

....

.... [3]

Total Marks .... / 5

Predicting Chemical Reactions

1 Which of the following correctly explains the reactivity of **Group 1** metals?

A   As the outermost electron is closest to the nucleus it is easier to remove that electron.

B   As the outermost electron is furthest from the nucleus it is easier to remove that electron.

C   The outermost electrons are closest to the nucleus and are easier to remove.

D   The outermost electrons are closer to the nucleus and are harder to remove.

Your answer ☐ [1]

2 Argon is a noble gas. It has the atomic number 18.

Which of the following correctly shows the electronic structure of argon?

A   B   C   D

Your answer ☐ [1]

3 Give **two** ways in which transition metals differ from metals in Groups 1 and 2.

....

.... [2]

4 Fluorine and oxygen are next to each other in the periodic table in Period 2.

Describe the similarities and differences between the reactivity of fluorine and oxygen.

....

....

....

....

.... [3]

Total Marks .... / 7

Identifying the Products of Chemical Reactions

1 Becky reacts a chemical, **X**, with hydrochloric acid.
A gas is formed which Becky then tests using limewater.
The limewater turns a milky white colour.

Which of the following is the gas that has been produced?

A hydrogen

B methane

C oxygen

D carbon dioxide

Your answer ☐ [1]

2 Describe the test for hydrogen gas.

....

.... [2]

3 A mystery ionic compound, **Z**, is tested with sodium hydroxide solution.
A **light blue** precipitate is observed.

a) What is the name of the **cation** detected?

.... [1]

b) The compound **Z** is then reacted with a few drops of hydrochloric acid and barium chloride solution.
A **white** precipitate appears.
The precipitate remains after adding more hydrochloric acid.

What is the identity of the **anion**?

.... [1]

c) Write the **chemical formula** for compound **Z**.

.... [1]

d) An iron nail is added to a test tube containing compound **Z**.

Describe what would be **seen** in the test tube.

....

.... [2]

Total Marks .... / 8

Ion Tests and Instrumental Methods of Analysis

1 Describe the procedure to detect whether a halide (halogen) ion is present in a solution.

[3]

2 Describe how to carry out a flame test.

[4]

3 Copper nitrate is reacted with sodium hydroxide.

Write the formulae of the ions that are reacting in this reaction.

[2]

4 An acid is tested with silver nitrate solution. A **cream** precipitate forms.

Which ion is present in the acid?

A silver   **B** bromide   **C** chloride   **D** sulfate

Your answer [ ] [1]

5 a) In a flame test, what is the reason for dipping the wire into concentrated hydrochloric acid?

[1]

b) What colour flame would you observe using the following ions?

   i) copper ions   [1]

   ii) calcium ions   [1]

   iii) potassium ions   [1]

Total Marks _____ / 14

Monitoring Chemical Reactions

1 Alice wants to find out how much hydrochloric acid is needed to neutralise sodium hydroxide.

a) Which process should she use?

    A chromatography   **B** filtration   **C** titration   **D** distillation

    Your answer ☐ [1]

b) Alice follows these steps:

    1. Measure the alkali into a conical flask using a pipette and filler.
    2. Add a few drops of indicator to the conical flask.
    3. Fill the burette with acid.

    What **two** steps are missing from her method? Write down the two letters.

    A Record the start volume of the acid.

    B Add alkali slowly until the indicator just changes colour.

    C Add acid slowly until the indicator just changes colour.

    D Pour acid into the conical flask.

    [2]

c) Suggest **two** ways Alice could improve the accuracy of her results.

    [2]

2 **a)** What is meant by mol/dm^3?

    [1]

b) Hydrochloric acid, HCl, has a molar mass of 36.5.

    If 109.5g HCl was dissolved in $1dm^3$ of water, what would the concentration be? Show your working.

    concentration = _____ [3]

Total Marks _____ / 9

Calculating Yields and Atom Economy

1. Hugh is making copper sulfate.
   Copper oxide reacts with sulfuric acid to form copper sulfate and water.

   The balanced symbol equation is: $CuO(s) + H_2SO_4(aq) \rightarrow CuSO_4(aq) + H_2O(l)$

   a) What is the **molar mass** of CuO to the nearest whole number?

   A 16   **B** 32   **C** 64   **D** 80

   Your answer ☐ [1]

   b) If there was 160g of CuO, how much $CuSO_4$ would be formed?

   $CuSO_4$ = ... [3]

   c) The answer to **b)** is a theoretical yield.

   What is meant by the term **theoretical yield**?

   .... [1]

   d) The amount of $CuSO_4$ formed was 160g.

   Calculate the percentage yield. Show your working.

   percentage yield = ... [3]

   e) Hugh needs to make 640g of copper sulfate.

   Given the yield in **d)**, calculate how much copper oxide he needs to react to get 640g $CuSO_4$.

   amount of copper oxide needed = ... [3]

   Total Marks .... / 11

Controlling Chemical Reactions

1 Which of the following shows the correct units for measuring the rate of reaction?

   A g/s   **B** g/m   **C** m/s   **D** cm^2/s

Your answer ☐ **[1]**

2 Which of the following correctly shows the reason why a chemical reaction stops?

   A   The maximum quantity of reactant has been made.

   B   The minimum quantity of reactant has been made.

   C   A reactant is used up.

   D   A product is used up.

Your answer ☐ **[1]**

3 Give **two** ways in which the rate of a reaction can be changed.

....

.... **[2]**

4 Look at the following diagrams.

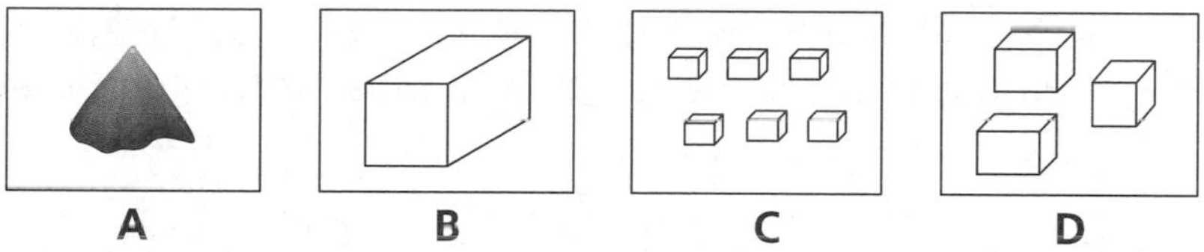

| **A** | **B** | **C** | **D** |

The diagrams all show the same mass of calcium carbonate.
Calcium carbonate reacts with nitric acid.

  a) Which diagram shows the calcium carbonate that would react the **slowest** with nitric acid?

Your answer ☐ **[1]**

  b) Describe how surface area affects the rate of reaction.

....

.... **[2]**

Total Marks _____ / 7

Catalysts and Activation Energy

1 Fraser investigated the reaction between copper carbonate and nitric acid.

He used the same amount of copper carbonate and nitric acid.

He changed the temperature.

He measured the amount of carbon dioxide produced and drew a graph.

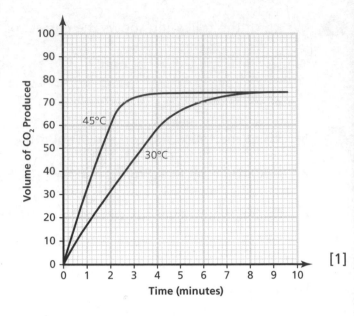

a) What does a **steeper** line on the graph indicate?

.... [1]

b) In Fraser's experiment, which **temperature** led to the quickest CO_2 production?

.... [1]

c) What was the **maximum** amount of CO_2 produced?

.... [1]

d) At what **time** did the reaction end for the experiment at 30°C?

.... [1]

e) Predict how long it would take for the reaction to finish if the temperature was 60°C.

.... [1]

2 Explain how a catalyst affects the rate of reaction.

....

....

....

.... [4]

Total Marks .... / 9

Equilibria

1 What is meant by the term **irreversible** reaction?

.... [1]

2 Which of the following reactions are classed as being **reversible**? Write down the correct letters.

A $CH_4(g) + 2O_2(g) \rightarrow CO_2(g) + 2H_2O(l)$

B $2Li(s) + 2HCl(aq) \rightarrow 2LiCl(aq) + H_2(g)$

C $H_2(g) + I_2(g) \rightleftharpoons 2HI(g)$

D $NH_4Cl(s) \rightleftharpoons NH_3(g) + HCl(g)$

.... [1]

3 The graph on the right shows a dynamic equilibrium.

Which of the following shows the correct labels for the graph?

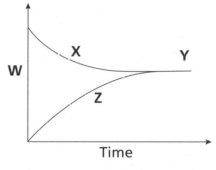

A W = forward reaction   X = reaction rate   Y = equilibrium   Z = backward reaction

B Y = forward reaction   Z = reaction rate   W = equilibrium   X = backward reaction

C X = forward reaction   W = reaction rate   Y = equilibrium   Z = backward reaction

D X = forward reaction   Y = reaction rate   W = equilibrium   Z = backward reaction

Your answer ☐ [1]

4 Look at the following reaction:

$H_2(g) + F_2(g) \rightleftharpoons 2HF(g)$

The pressure of the system is **decreased**.

How will this affect the position of the equilibrium?

.... [1]

Total Marks .... / 4

Improving Processes and Products

1 What is the name given to a rock containing minerals of metal compounds?

_____ [1]

2 Which of the following is the element that most metals on the planet are bonded to?

A sulfur   B silica   C oxygen   D carbon

Your answer ☐ [1]

3 Andy is carrying out the experiment shown in the diagram below.

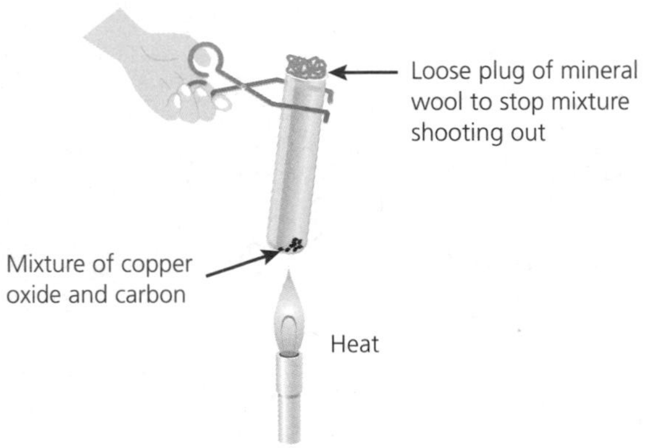

Loose plug of mineral wool to stop mixture shooting out

Mixture of copper oxide and carbon

Heat

a) The loose plug of mineral wool in the top of the test tube is a safety precaution.

Give **two** other safety precautions that should be carried out when doing the experiment.

_____

_____ [2]

b) Andy reacts copper(II) oxide with carbon.

Write a **word** equation for the reaction that takes place.

_____ [2]

c) Cuprite, an ore of copper, has the formula Cu_2O.

Write a **balanced symbol** equation for the reaction between cuprite and carbon.

_____ [2]

Total Marks _____ / 8

The Haber Process

1 The Haber process and the contact process are used to make ammonium sulfate.

Look at the flowchart.

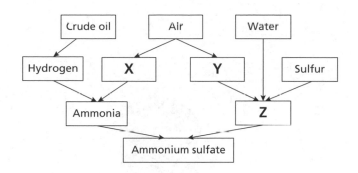

Which of the following shows the correct labels for **X**, **Y** and **Z**?

A X = oxygen   Y = nitrogen   Z = ammonia

B X = oxygen   Y = nitrogen   Z = sulfuric acid

C X = nitrogen   Y = oxygen   Z = ammonia

D X = nitrogen   Y = oxygen   Z = sulfuric acid

Your answer ☐ [1]

2 Write the **word** equation for the Haber process.

.... [2]

3 It is important that the maximum amount of ammonia is made in the shortest possible time at a reasonable cost.

Explain why the Haber process is carried out:
- at a temperature of 450°C
- at a pressure of 200 atmospheres
- with an iron catalyst.

....

....

[4]

Total Marks .... / 7

Life Cycle Assessments , Recycling and Alloys

1 Lochmara Lodge Café prides itself on being environmentally friendly.
Inside the café the lights have lampshades made out of used computer circuit boards.

The lampshades are an example of recycling.

Give **three** reasons why recycling is beneficial.

....

....

....

....

[3]

2 Many of the metals used to make products are alloys.
One of the most used alloys is steel.
Steel is a mixture of the elements carbon and iron.

Explain, with the aid of a diagram, how the addition of carbon to iron makes the steel alloy much stronger and less liable to shearing.

....

....

....

....

....

[4]

Total Marks .... / 7

Using Materials

1 Mark is an artist and has been commissioned to build a metal sculpture that will be displayed in London's Trafalgar Square. He is deciding which material to make the sculpture from.

Material	Density (g/cm³)	Strength	Corrosion Resistance	Availability
Steel	7.85	Medium	Low	High
Copper	8.96	Low	High	Medium
Zinc	7.13	Low	Medium	Low
Bronze	8.51	High	Very High	Medium

Mark wants to make a statue that will last for the next 100 years.

a) Which factor is the most important for this requirement?

    A density   **B** strength   **C** corrosion resistance   **D** availability

                                              Your answer [ ] **[1]**

b) Iron rusts easily.

Write the **word** equation for the rusting of iron.

    .... **[2]**

c) Mark is told that the cost of bronze is too high.
He has been told to make the sculpture out of steel.

Explain what Mark will have to do to ensure that his sculpture will last 100 years.

    ....

    .... **[3]**

2 Match the polymer properties to their uses by drawing a line to the correct use.

Poly(ethene) – light and flexible		insulation

Poly(styrene) – light and poor heat conductor		moulded containers

Polyester – tough and can be coloured		clothing

                                                          [2]

Total Marks .... / 8

Organic Chemistry

1　**a)** Which of the following is an **alkane**?

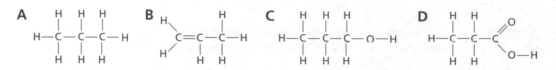

　　　Your answer ☐ [1]

b) Each of the molecules **A–D** has three carbon atoms.

Which of the following shows the correct prefix for three carbons?

A Meth　　　　**B** Eth　　　　**C** Prop　　　　**D** But

　　　Your answer ☐ [1]

c) Write the **structural formula** for butanoic acid in the space below.　　[2]

d) What is meant when a molecule is said to be **unsaturated**?

.... [1]

e) Which of the molecules **A–D** are **unsaturated**?

.... [1]

f) Write the functional group for an **alcohol**.

.... [1]

g) Suggest what group of organic chemicals this molecule belongs to.

.... [1]

Total Marks _____ / 8

Organic Compounds and Polymers

1 Farah has a mystery chemical, **X**.
She bubbles a small amount into orange-coloured bromine water.
The bromine water is decolourised.

What conclusion can Farah draw from this finding?

A **X** must be an alcohol.   **B** **X** must be an alkene.

C **X** must be an alkane.   **D** The bromine water is not working correctly.

Your answer ☐ [1]

2 The diagram shows the displayed formula for propene.

Propene is reacted with hydrogen gas.

Write the **word** equation for this reaction.

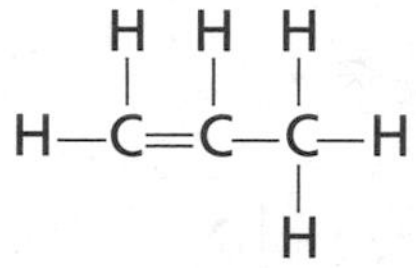

$$H-\overset{\displaystyle H}{\underset{}{C}}=\overset{\displaystyle H}{\underset{}{C}}-\overset{\displaystyle H}{\underset{\displaystyle H}{C}}-H$$

.... [1]

3 The following monomer units undergo a condensation reaction:

$$OH-\overset{\displaystyle O}{\overset{\displaystyle \|}{C}}-\square-\overset{\displaystyle O}{\overset{\displaystyle \|}{C}}-OH \qquad H_2N-\square-NH_2$$

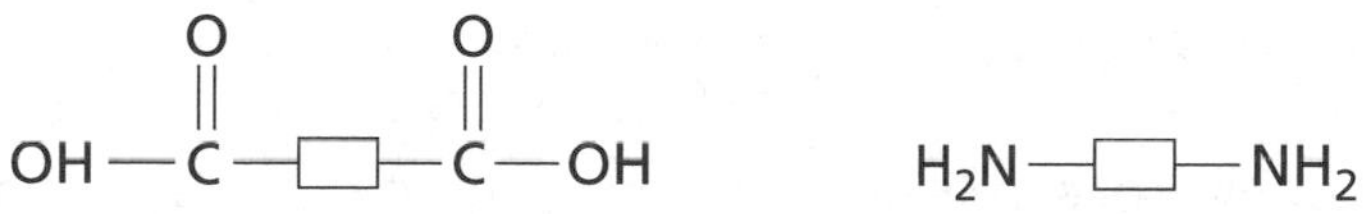

Which of the following shows the correct products from this condensation reaction?

A

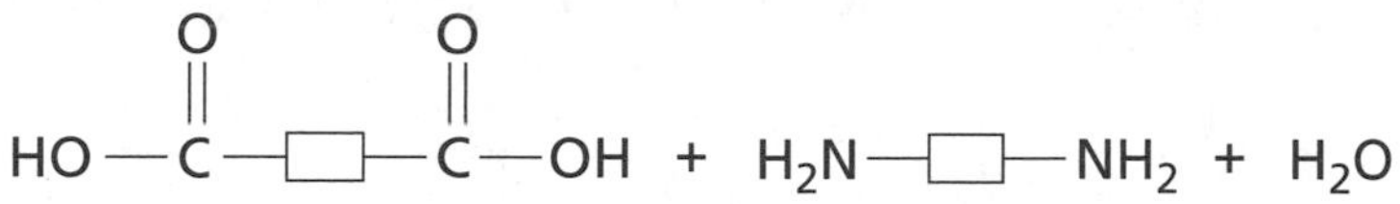

$$HO-\overset{\displaystyle O}{\overset{\displaystyle \|}{C}}-\square-\overset{\displaystyle O}{\overset{\displaystyle \|}{C}}-OH \ + \ H_2N-\square-NH_2 \ + \ H_2O$$

B

$$\left[\overset{\displaystyle O}{\overset{\displaystyle \|}{C}}-\square-\overset{\displaystyle O}{\overset{\displaystyle \|}{C}}-\overset{\displaystyle H}{\overset{\displaystyle |}{N}}-\square-\overset{\displaystyle H}{\overset{\displaystyle \|}{N}}\right]_n$$

C

$$\left[\overset{\displaystyle O}{\overset{\displaystyle \|}{C}}-\square-\overset{\displaystyle O}{\overset{\displaystyle \|}{C}}-\overset{\displaystyle H}{\overset{\displaystyle |}{N}}-\square-\overset{\displaystyle H}{\overset{\displaystyle \|}{N}}\right]_n \ + \ nH_2O$$

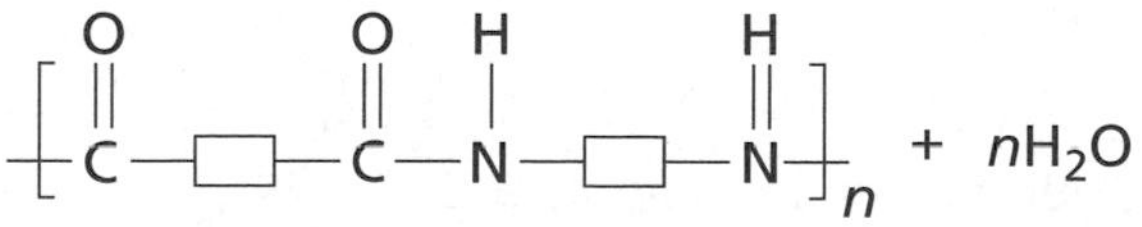

D

$$\left[\overset{\displaystyle O}{\overset{\displaystyle \|}{C}}-\square-\overset{\displaystyle O}{\overset{\displaystyle \|}{C}}-\overset{\displaystyle H}{\overset{\displaystyle |}{N}}-\square-\overset{\displaystyle H}{\overset{\displaystyle |}{N}}\right]_n \ + \ NH_3$$

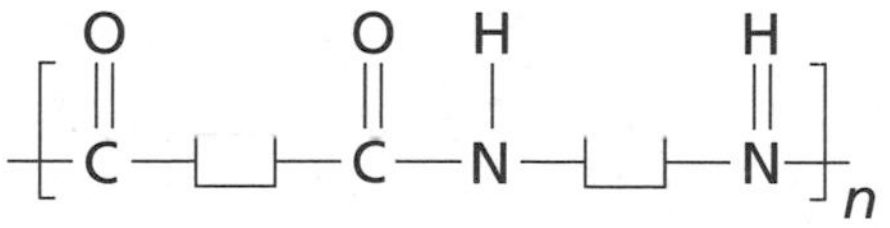

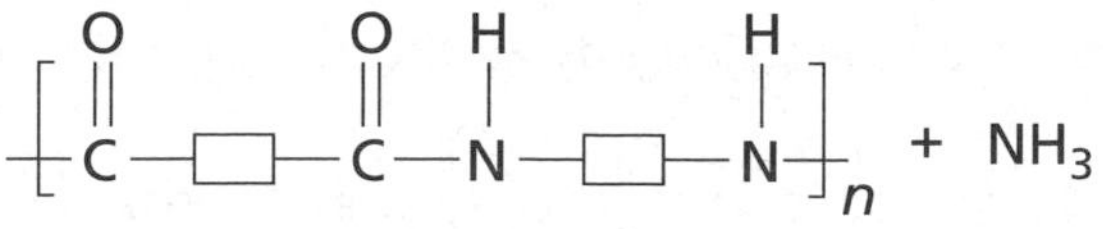

Your answer ☐ [1]

Total Marks .... / 3

Crude Oil and Fuel Cells

1 The diagram shows a fractional distillation column.

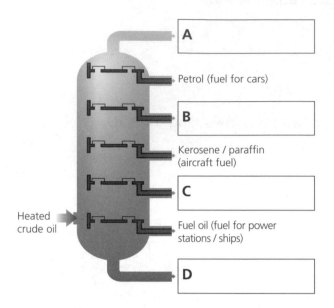

A

Petrol (fuel for cars)

B

Kerosene / paraffin
(aircraft fuel)

C

Heated
crude oil

Fuel oil (fuel for power
stations / ships)

D

The diagram is missing the following labels:

Bitumen   Diesel   Naphtha   Refinery gases

Write each label in the correct place on the diagram. [4]

2 Explain why short chain hydrocarbons have a low boiling point and long chain hydrocarbons have a high boiling point.

....

.... [2]

3 Not all of the products from fractional distillation are useful.
To make them useful, they have to be cracked.

What is meant by the term **cracking**?

.... [1]

4 In 2015, BOC announced a large, portable hydrogen cell called Hymera II™.
It can produce 150W of d.c. current for off-grid and backup power applications.
It has a mass of 20kg, but produces the same amount of electricity as 200kg of car batteries.

What are two **advantages** of using a hydrogen cell, compared to getting energy from crude oil?

.... [2]

Total Marks _____ / 9

Interpreting and Interacting with Earth's Systems

1 The atmosphere of our Earth currently is made up of the following gases.

Gas	Proportion of the Atmosphere (%)
Nitrogen	78
Oxygen	21
Carbon dioxide	0.04
Other gases	1

a) The numbers in the table do not add up to 100%.

Suggest why this is the case.

.... [1]

b) Scientists have created models of the atmosphere of early Earth.
One of the models is shown below. It is in the **incorrect** order.

Put these statements into the correct order.

A These newly formed oceans removed some carbon dioxide by dissolving the gas.

B Primitive plants that could photosynthesise removed carbon dioxide from the atmosphere, and added oxygen.

C Levels of nitrogen in the atmosphere increased as nitrifying bacteria released nitrogen.

D A hot volcanic Earth released ammonia, carbon dioxide and water vapour.

E The Earth cooled and water vapour condensed into liquid water.

The **correct** order is:

[2]

2 What part of the electromagnetic spectrum is affected by greenhouse gases?

A Gamma   **B** Ultraviolet   **C** Infrared   **D** Heat

Your answer ☐ [1]

Total Marks .... / 4

1 Draw a line from each air pollutant to the effect it produces.

Carbon monoxide	coats surfaces with soot
Sulfur dioxide	kills plants and aquatic life
Nitrogen oxides	forms photochemical smogs
Particulates	air poisoning

[3]

2 Clean drinking water is essential for everyday life.
Look at the flowchart.
It shows the treatment of sewage water leading to fresh drinking water.

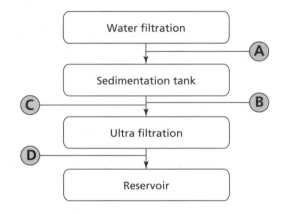

a) What is **removed** from the water treatment system at **A** and **B**?

A ....

B .... [2]

b) What is **added** to the water treatment system at **C** and **D**?

C ....

D .... [2]

3 Draw a diagram to show how **thermal desalination** takes place.

[2]

Total Marks .... / 9

Collins

GCSE (9–1)
Chemistry
Paper 1 Higher Tier

Time allowed: 1 hour 45 minutes

You must have:
- the Data Sheet (page 224)

You may use:
- a scientific calculator
- a ruler

Instructions

- Use black ink. You may use an HB pencil for graphs and diagrams.
- Answer **all** the questions.
- Write your answer to each question in the space provided.
- Additional paper may be used if required.

Information

- The total mark for this paper is **90**.
- The marks for each question are shown in brackets **[ ]**.
- Quality of extended responses will be assessed in questions marked with an asterisk (*).

SECTION A

Answer **all** the questions.

You should spend a maximum of 30 minutes on this section.

1   Athina wants to separate a mixture of amino acids.

She carries out thin layer chromatography.

The amino acid samples are placed onto silica gel and a solvent is added.

After two hours the silica gel is sprayed with ninhydrin, which makes the amino acids become visible.

What is the name given to the **silica gel** in this experiment?

A   gas phase

B   mobile phase

C   solid phase

D   stationary phase

Your answer ☐   [1]

2   The diagram below shows the chromatogram before and after spraying with ninhydrin.

What is the R_f value for sample **3**?
Use a ruler to help you.

A   0.16

B   0.27

C   0.81

D   0.95

Your answer ☐   [1]

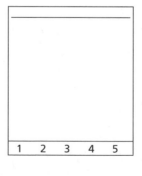
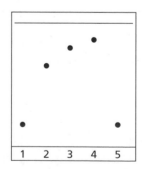

Before   After

3   The diagrams show the electronic structure of four different atoms.

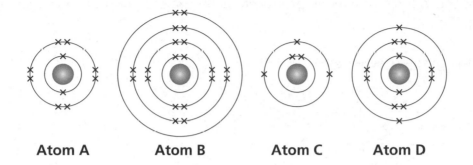

Atom A   **Atom B**   **Atom C**   **Atom D**

Which two atoms are in the same **group**?

A   atom A and atom B

B   atom B and atom D

C   atom C and atom D

D   atom A and atom C

Your answer ☐   **[1]**

4   An element, **X**, has an atomic mass of 88 and the atomic number 38.

Which of the following is correct for element **X**?

A   No. of protons = 38   No. of neutrons = 50   No. of electrons = 50

B   No. of protons = 50   No. of neutrons = 38   No. of electrons = 38

C   No. of protons = 38   No. of neutrons = 38   No. of electrons = 50

D   No. of protons = 38   No. of neutrons = 50   No. of electrons = 38

Your answer ☐   **[1]**

Turn over

5 Wayne is producing oxygen.

He measures the volume of oxygen produced by 50 cm³ of hydrogen peroxide in the presence of a catalyst.

He obtains the graph shown below.

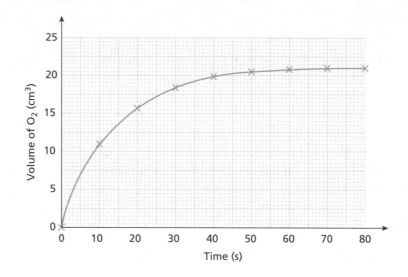

Wayne had calculated that he should have collected 30 cm³ of oxygen.

Which of the following correctly shows the percentage yield in this experiment?

A 21%

B 42%

C 30%

D 70%

Your answer ☐ [1]

6 During the electrolysis of aqueous copper sulfate, what is made at the cathode?

   A sulfate

   B hydrogen

   C copper

   D copper hydroxide

   Your answer ☐   **[1]**

7 Which of these shows the balanced symbol equation for the reaction between sodium and oxygen to make sodium oxide?

   A $Na + O_2 \rightarrow NaO_2$

   B $4Na + O_2 \rightarrow 2Na_2O$

   C $S + O_2 \rightarrow SO_2$

   D $4S + O_2 \rightarrow 2S_2O$

   Your answer ☐   **[1]**

8 The molecular formula of octadecane is $C_{18}H_{38}$.

   What is the empirical formula of octadecane?

   A CH_2

   B C_2H_4

   C C_9H_{19}

   D $C_{36}H_{76}$

   Your answer ☐   **[1]**

9   What is the best description of the particles in a solid?

	Distance between particles	Movement of particles
A	Close together	In continuous random motion
B	Close together	Vibrating about a fixed point
C	Far apart	In continuous random motion
D	Far apart	Vibrating about a fixed point

Your answer [ ]   [1]

10   Saritha neutralises sulfuric acid with sodium hydroxide solution.

Which of these shows the net ionic equation for the reaction?

A $H_2SO_4 + 2NaOH \rightarrow Na_2SO_4 + 2H_2O$

B $H^+ + OH^- \rightarrow H_2O$

C $SO_4^{2-} + 2Na^+ \rightarrow Na_2SO_4$

D $2H^+ + SO_4^{2-} \rightarrow H_2SO_4$

Your answer [ ]   [1]

11   Spencer is investigating some acids.

He has a solution of nitric acid at a concentration of 0.001 mol/dm^3.

This solution has a pH of 3.

He increases the concentration of nitric acid by a factor of 100 from 0.001 mol/dm^3 to 0.1 mol/dm^3.

What is the pH of the new solution?

A   1

B   2

C   0

D   13

Your answer [ ]   [1]

12 Which of these is the **best** explanation of what is meant by a weak acid?

    A There is a large amount of acid and a small amount of water.

    B There is a small amount of acid and a large amount of water.

    C The acid is completely ionised in solution in water.

    D The acid is partially ionised in solution in water.

Your answer ☐ **[1]**

13 Limestone is the common name for calcium carbonate.

Limestone reacts with hydrochloric acid.

Which of the following shows the correct balanced symbol equation for the reaction between limestone and hydrochloric acid?

    A $CaCO_3 + HCl \rightarrow CaCl + H_2O$

    B $CaCO_3 + HCl \rightarrow CaCl + H_2O + CO_2$

    C $CaCO_3 + 2HCl \rightarrow CaCl_2 + H_2O + CO_2$

    D $CaCO_3 + 2HCl \rightarrow CaCl_2 + H_2O + CO_3$

Your answer ☐ **[1]**

14 Which of the following shows the reaction profile for an exothermic reaction?

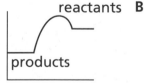
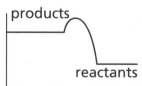
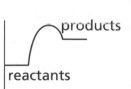
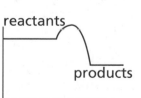

Your answer ☐ **[1]**

Turn over

15 The table shows some properties of alkanes at 25°C.

Name	Molecular formula	Molecular mass	Alkanes at room temperature
Methane	CH_4	16	Gas
Butane	C_4H_{10}	58	Gas
Hexane	C_6H_{14}	86	Liquid
Decane	$C_{10}H_{22}$	142	Liquid
Eicosane	$C_{20}H_{42}$	282	Solid

Using the information in the table, which of the following is most likely to be **true**?

A As the molecular mass increases, there are more intermolecular forces between chains. The alkanes will be gases for molecular masses below 85.

B As the molecular mass decreases, there are more intermolecular forces between chains. The alkanes will be solids for molecular masses above 280.

C As the molecular mass increases, there are more intermolecular forces between chains. The alkanes will be liquid for molecular masses above 280.

D As the molecular mass decreases, there are fewer intermolecular forces between chains. The alkanes will be gases for molecular masses below 280.

Your answer ☐

[1]

SECTION B

Answer **all** the questions.

16 Look at the graph of boiling points of elements against their atomic numbers for the second period of the periodic table.

(a) What does the graph tell you about the boiling point as you go across a **period**?

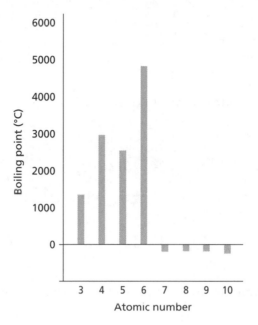

....

.... [2]

(b) Explain why the properties of elements change across a period.

....

....

....

.... [2]

Turn over

(c) Particle **X** is a particle with the atomic number of 18 and the electronic structure 2.8.8.

Particle **Y** is a particle with the atomic number of 20 and the electronic structure 2.8.8.

Which particle is an atom and which is an ion?

Explain your answer.

[2]

(d) A diagram of a hydrogen atom is shown below.

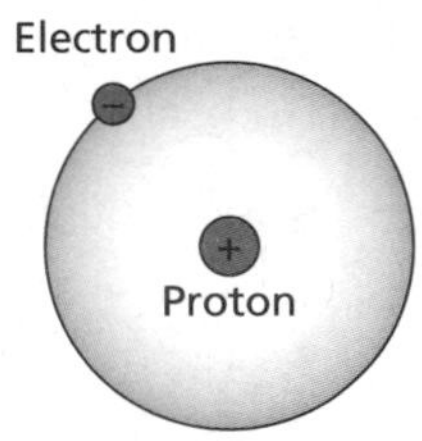

Electron

Proton

(i) Explain what the **problems** are in representing the atom in this way.

[3]

(ii) Niels Bohr was the scientist who devised this particular model of the atom.

Explain how his model **improved** upon earlier models.

[2]

17 The diagram below shows an allotrope of carbon called graphene.

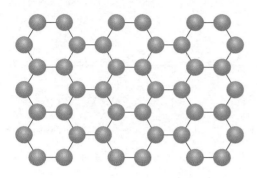

(a) Graphene is used in solar panels.

Explain what **properties** make graphene useful for solar panels.

....

.... **[2]**

(b) Another form of carbon is graphite.

As well as being a good conductor of electricity it can also be used as a dry lubricant.

Draw the structure of graphite in the space below. **[2]**

(c) Diamond and graphite have been known for thousands of years.

Graphene was discovered in 2004 whilst another form of carbon, the fullerenes, were discovered in 1985.

Suggest why graphene and the fullerenes were only discovered relatively recently.

....

.... **[2]**

Turn over

18 **(a)** Look at the reaction profile.

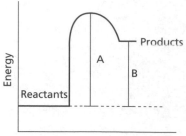

Explain, with reasons, what you can deduce about this reaction.

....

....

....

....

....

....

[4]

(b) Look at the equation. It shows the combustion of hydrogen.

$$H-H$$

$$+ \quad O=O \quad \rightarrow \quad \begin{matrix} & O & \\ H & & H \\ & O & \\ H & & H \end{matrix}$$

$$H-H$$

Hydrogen   Oxygen   Water

The table shows the bond energies of the bonds involved.

Bond	Bond energy (kJ/mol)
H–H	432
O=O	494
O–H	464

(i) Write down the:

number of bonds broken = ...

number of bonds made   = ...   **[1]**

(ii) Describe the energy changes when bonds are broken and made.

....

.... **[2]**

(iii) Calculate the energy change for this reaction. Show your working.

energy change = ... kJ/mol   **[3]**

Turn over

19 Look at the diagram. It shows the electrolysis of copper sulfate solution.

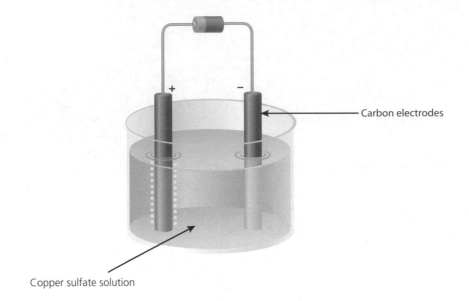

Carbon electrodes

Copper sulfate solution

(a) Explain what happens to the copper ions during electrolysis.

[3]

(b) Write the half equation for the reaction at the cathode.

[2]

(c) Oxygen is produced at the carbon electrode.

What would you expect to see?

[2]

(d) Describe how you could change this experiment so that you could silver plate an object, such as a metal earring.

[2]

20   Niles is identifying a white solid, **X**.

He dissolves **X** in water to give a colourless solution.

(a) Niles decides to carry out a flame test.

He suggests the following method for carrying out the flame test.

> 1. Take a wire loop and dip in water to clean.
> 2. Dip the loop into the solution of **X**.
> 3. Insert the tip into the Bunsen flame.
> 4. Note the colour of the flame.

What improvements should be made to Niles's method to ensure that he sees the greatest amount of colour in the flame?

Explain your answer.

....

....

....

....   **[4]**

(b) On carrying out the flame test with substance **X**, he observed a crimson colour.

Name the ion present in substance **X**.

....   **[1]**

(c) Niles next adds some dilute nitric acid and silver nitrate to the solution of substance **X**.

A cream precipitate appears.

(i) What is the name of the anion that was present in substance **X**?

....   **[1]**

(ii) Write the balanced ionic equation for this reaction.

....   **[1]**

(d) Draw the structure of the solid form of **X**.

[2]

Turn over

21 In the food industry, nanoparticles are being introduced to food packaging.

They can detect bacteria that will spoil the food.

(a) What is the size range of a particle to be classified as being a nanoparticle?

.... [1]

(b) Explain how the properties of nanoparticles make them suitable for their use in detecting food spoilage.

....

....

....

[2]

(c) Food spoilage makes a number of people ill every year and some people may die from food poisoning.

Even with the benefits, some people are concerned about the use of nanoparticles.

Discuss the risks involved with using nanoparticles embedded into the plastic packaging of food.

....

....

....

....

[3]

22 A student is investigating the reaction of three acids, **X**, **Y** and **Z**.

Acid	pH	Dissociation Constant
X	1	7.6×10^{-4}
Y	2	1.8×10^{-5}
Z	5	8.7×10^{-10}

(a)*Suggest what the dissociation constant means in relation to acids **X**, **Y** and **Z** and explain in detail the differences between the different types of acid.

Explain how you could test the strengths of the acids experimentally.

....

....

....

....

....

....

....

....

....

....

....

....

.... **[6]**

(b) How much stronger is acid **X** compared to acid **Y**?

.... **[1]**

Turn over

23 Melinda is making a compound.

She reacts magnesium with a diatomic liquid, **X**, to form a single product.

2.43 g of magnesium reacts with 15.98 g of **X**.

(a) (i) What mass of magnesium **X** is made?

.... **[1]**

(ii) Calculate the number of moles of magnesium, **X** and magnesium **X** involved in the reaction.

(The relative formula mass of magnesium, Mg = 24.3, **X** = 79.9 and MgX$_2$ = 184.1)

number of moles of magnesium = ...

number of moles of **X** = ...

number of moles of magnesium **X** = ... **[3]**

(b) Use your answers to write the **balanced symbol** equation for the reaction between magnesium and **X**.

.... **[2]**

(c) Give the formula of the ions present in MgX$_2$

.... **[1]**

24 A teacher carries out a thermite reaction.

The thermite reaction is shown below.

aluminium + iron(III) oxide → iron + aluminium oxide

(a) Write a **balanced symbol** equation to show this reaction.

.... [1]

(b) Suggest **two** safety precautions a person carrying out this reaction should take.

....

....

.... [2]

(c) The thermite reaction is an example of an oxidation / reduction reaction.

Write the formula of the species that is:

oxidised ....

reduced .... [2]

(d) Calculate the mass of iron(III) oxide that has to be reacted to make 10.2 g of aluminium oxide.

Give your answer to three significant figures.

The relative atomic mass of Fe = 55.8, O = 16.0, Al = 27.0.

mass = ... g   [3]

Turn over

25 Stevioside, or Stevia, is a sweetener that is a natural alternative to sugar (sucrose).

The relative molecular mass of Stevioside is 805.

(a) Calculate what the mass of a single molecule of Stevia would be.

Give your answer to three significant figures.

mass = _____ g   [1]

(b) Sucrose has the formula $C_{12}H_{22}O_{11}$.

The relative atomic mass of C = 12.0, H = 1.0 and O = 16.0.

How much heavier is a Stevia molecule when compared to a molecule of sucrose?

Give your answer to three significant figures.

answer = _____ **times heavier**   [1]

END OF QUESTION PAPER

Collins

GCSE (9–1)
Chemistry
Paper 2 Higher Tier

Time allowed: 1 hour 45 minutes

You must have:

- the Data Sheet (page 224)

You may use:

- a scientific calculator
- a ruler

Instructions

- Use black ink. You may use an HB pencil for graphs and diagrams.
- Answer **all** the questions.
- Write your answer to each question in the space provided.
- Additional paper may be used if required.

Information

- The total mark for this paper is **90**.
- The marks for each question are shown in brackets **[ ]**.
- Quality of extended responses will be assessed in questions marked with an asterisk (*****).

SECTION A

Answer **all** the questions.

You should spend a maximum of 30 minutes on this section.

1   Which of the following is a property of all Group 0 elements?

   A They have eight electrons in the outermost shell.

   B They are chemically unreactive.

   C They form positive ions.

   D They are less dense than air.

   Your answer [ ]   [1]

2   Look at the graph.

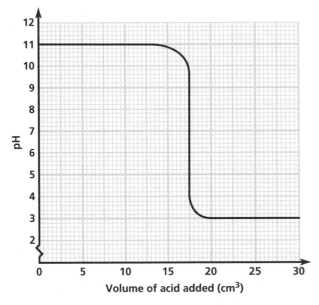

   Which of the following shows the volume of the titre at neutralisation in this titration?

   A 3.0 cm^3

   B 11.0 cm^3

   C 15.0 cm^3

   D 17.5 cm^3

   Your answer [ ]   [1]

3 Long hydrocarbon chains can be broken into smaller chain hydrocarbons.

Look at the following reaction:

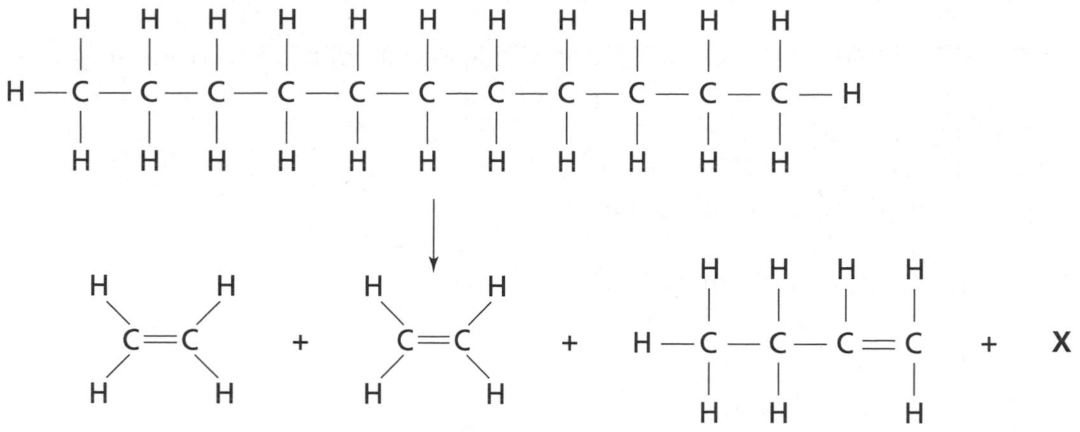

What is the identity of product **X**?

A ethene

B butane

C propane

D propene

Your answer ☐   [1]

4 Liquid phosphorus trichloride has the formula PCl_3.

A scientist makes 2 moles of PCl_3 from 3 moles of Cl_2.

Calculate the mass of PCl_3 made.

The relative atomic mass, A_r, of P = 31.0 and Cl = 35.5.

A 213.0 g

B 275.0 g

C 137.5 g

D 106.5 g

Your answer ☐   [1]

Turn over

5 Olwen is trying to determine the identity of a mystery compound.

She tests for the cation present in solution.

Which of the following is the best test to detect the cation in solution?

A Add acidified hydrochloric acid solution followed by barium chloride solution.

B Mix with silver nitrate solution.

C Displace using copper sulfate solution.

D Add a few drops of sodium hydroxide solution.

Your answer [ ]   [1]

6 Miranda is investigating the breakdown of hydrogen peroxide into water and oxygen.

$$2H_2O_2 \rightarrow 2H_2O + O_2$$

The reaction is extremely slow.

Miranda takes four powders which she adds to the hydrogen peroxide.

She records the amount of gas produced.

Look at the graph of her results.

Which of the catalysts tested would be the best for catalysing this reaction?

A copper oxide

B manganese dioxide

C tungsten dioxide

D zinc oxide

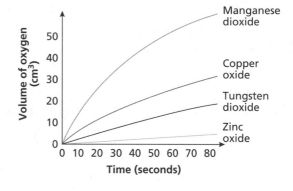

Your answer [ ]   [1]

7   One method of making methanol is to react a molecule called bromomethane with water.

The reaction for this is shown below:

$$CH_3Br + H_2O \rightarrow CH_3OH + HBr$$

Which of the following shows the atom economy for producing methanol, CH_3OH?

The relative atomic mass, A_r, of C = 12, H = 1, Br = 79.7 and O = 16.

A   28%

B   30%

C   40%

D   72%

Your answer [   ]   **[1]**

8   Alice wants to make a solution of copper sulfate using copper sulfate crystals.

Which of the following would be the best way for Alice to make a 0.5 mol/dm³ solution of copper sulfate?

The relative atomic mass, A_r, of Cu = 63.5, S = 32.1 and O = 16.0.

A   Dissolving 159.6 g of copper sulfate crystals in 1.0 dm³ of water.

B   Dissolving 79.8 g of copper sulfate crystals in 1.0 dm³ of water.

C   Dissolving 159.6 g of copper sulfate crystals in water to make 0.1 dm³ of solution.

D   Dissolving 79.8 g of copper sulfate crystals in water to make 0.1 dm³ of solution.

Your answer [   ]   **[1]**

9   Which of the following displayed formulae shows a compound with a **carboxylic acid group**?

A

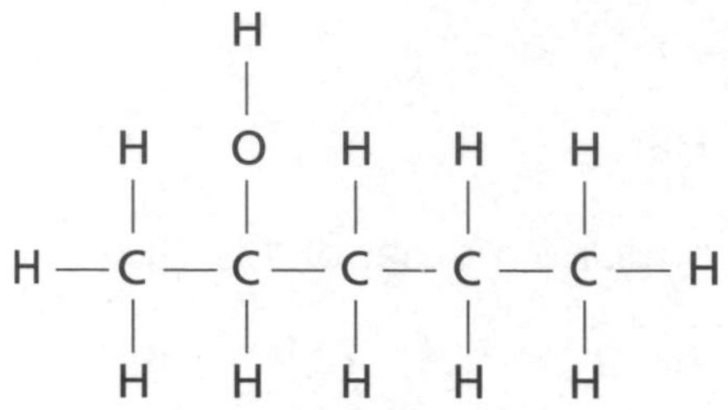

B

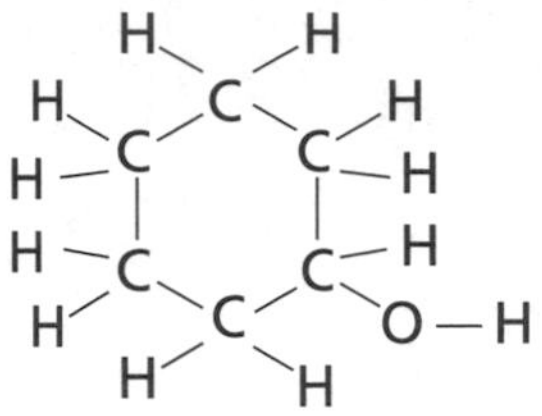

C

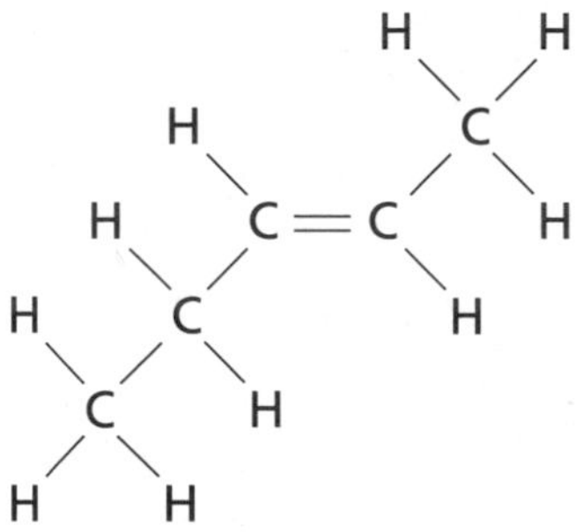

D

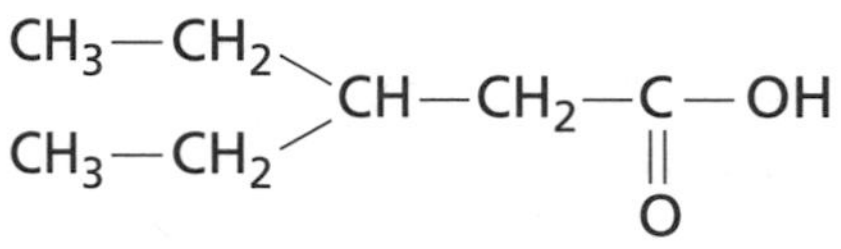

Your answer [ ]   [1]

10 A balloon used for long-distance flying contains 1.2×10^7 dm^3 of helium.

The molar volume, measured at room temperature and pressure, is 24 dm^3.

Assume the balloon is currently on the ground with 1 atm of air pressure at 25 °C.

Which of the following shows the number of moles and the number of atoms of helium in the balloon?

A No. of moles = 5.0×10^5 No. of atoms = 3.0×10^{29}

B No. of moles = 5.0×10^5 No. of atoms = 1.5×10^{29}

C No. of moles = 2.88×10^8 No. of atoms = 3.0×10^{29}

D No. of moles = 2.88×10^8 No. of atoms = 1.2×10^{21}

Your answer ☐   [1]

11 Molten aluminium oxide can be electrolysed to produce molten aluminium and oxygen gas.

Which of the following shows the balanced half equations at each electrode?

A Anode: $2O^{2-} \rightarrow O_2 + 4e^-$

   Cathode: $Al^{3+} + 3e^- \rightarrow Al$

B Anode: $2O^{2-} + 2e^- \rightarrow O_2$

   Cathode: $Al^{3+} \rightarrow Al + 3e^-$

C Anode: $O_2 + 2e^- \rightarrow 2O^{2-}$

   Cathode: $Al \rightarrow Al^{3+} + 3e^-$

D Anode: $Al^{3+} + 3e^- \rightarrow Al$

   Cathode: $2O^{2-} \rightarrow O_2 + 2e^-$

Your answer ☐   [1]

Turn over

12   Modern body armour is made of a polymer called Kevlar™.

Kevlar™ is made from the following monomer units:

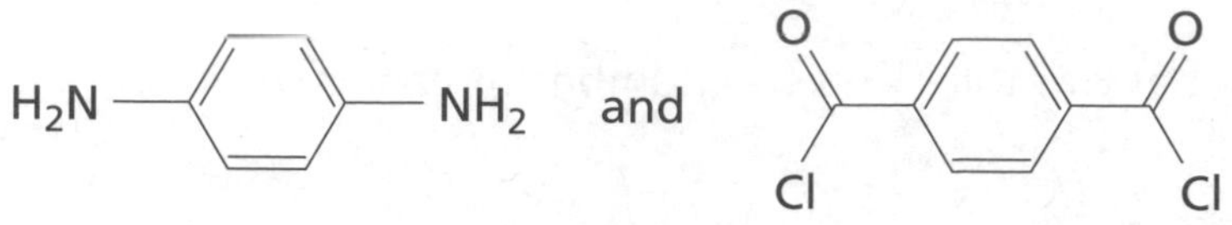

They undergo a **condensation polymerisation** reaction by the removal of hydrogen chloride molecules.

Which of the following shows the correct products from this condensation reaction?

A

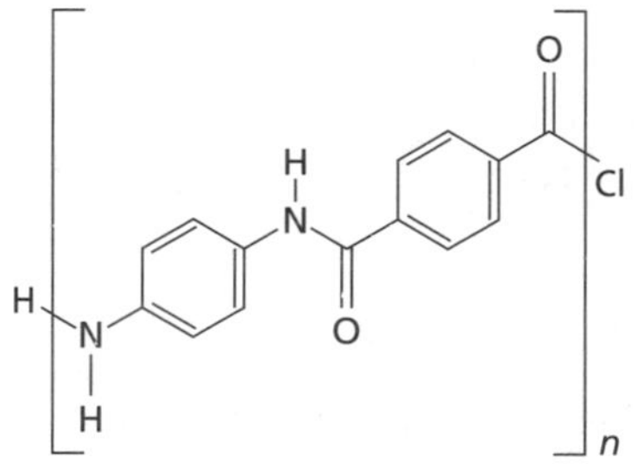

B

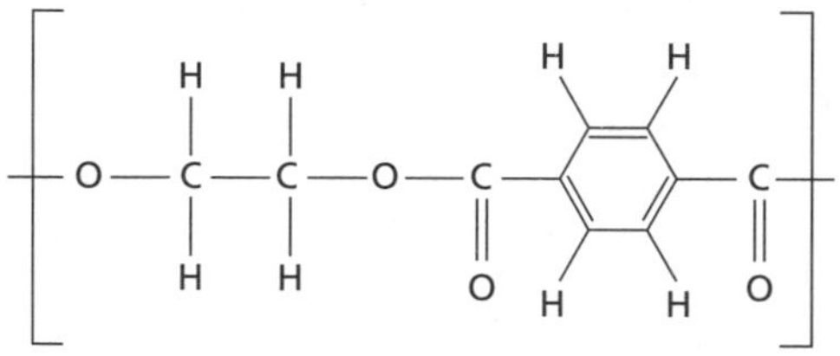

C

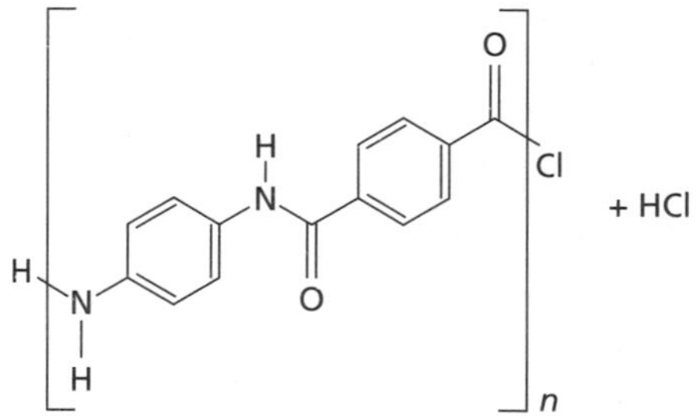

+ HCl

D

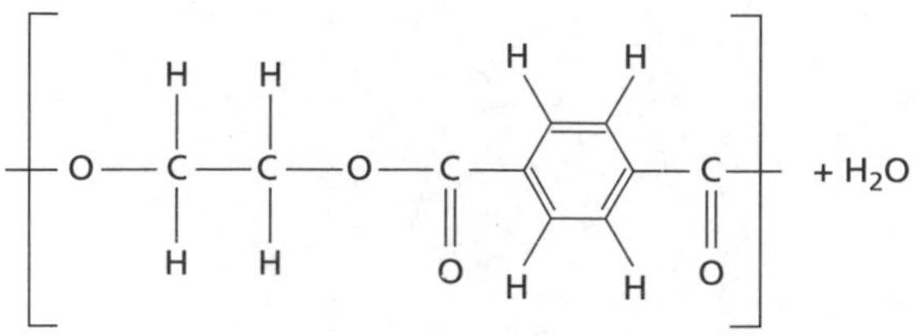

+ H_2O

Your answer ☐

[1]

13 A student bubbles propene into bromine water.

Which of the following displayed formulae shows the product for this reaction?

A

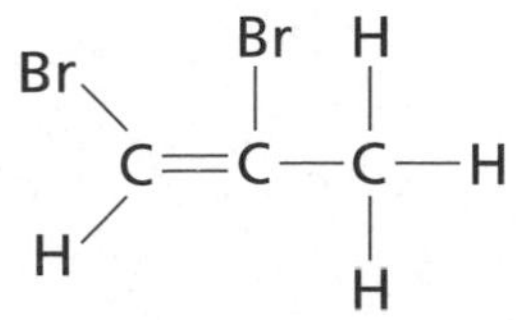

B

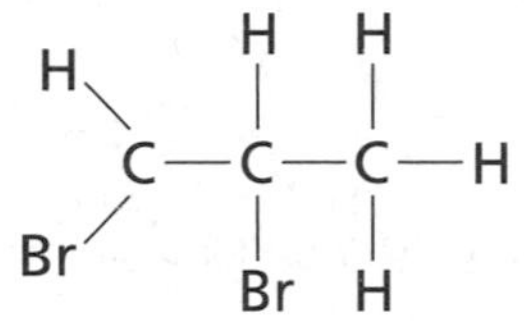

C

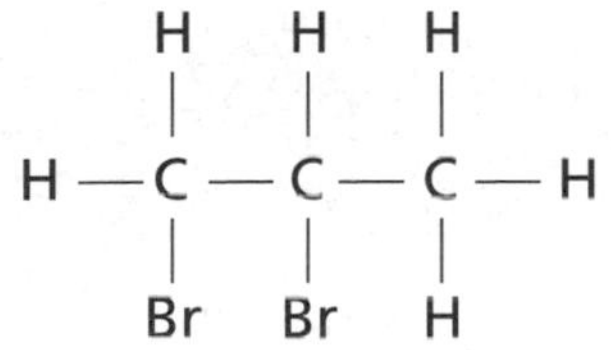

D

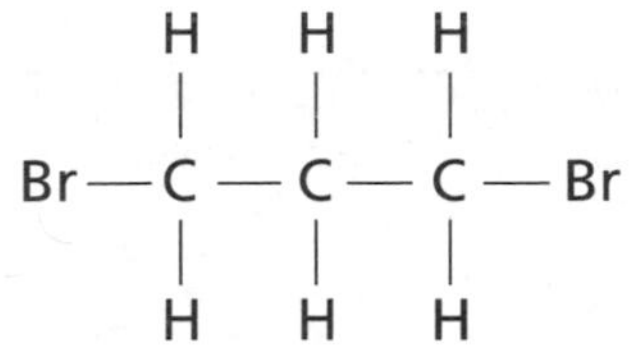

Your answer ☐ **[1]**

Turn over

14 Siobhan is carrying out tests on some organic molecules.

She adds a mystery molecule to acidified potassium manganate(VII) solution.

The product of the reaction is:

The table shows some results.

Which row shows the mystery molecule Siobhan used and the result of mixing with acidified potassium manganate(VII)?

	Mystery molecule	Colour of acidified potassium manganate(VII)	
		Before addition	After addition
A		Colourless	Purple
B		Purple	Colourless
C	H—C≡C—H	Colourless	Purple
D	H—C≡C—H	Purple	Colourless

Your answer ☐   [1]

15 Methane is an organic molecule found naturally in the atmosphere.

Which of the following is **not** true for methane?

A It is a greenhouse gas.

B It is produced by living things.

C It is more effective as a greenhouse gas than CO_2.

D Plants can take in methane when they photosynthesise.

Your answer ☐ **[1]**

SECTION B

Answer **all** the questions.

16 **(a)** Give the electronic structure of potassium.

.... [1]

(b) Draw a diagram to show the metallic bonding in potassium.

[2]

(c) A small piece of potassium is added to water.

Describe what observations you would make when a small piece of potassium is added to a trough of water.

....

....

.... [2]

(d) Write the word equation to show the reaction between water and potassium.

.... [2]

17 The table below shows the composition of the Earth's atmosphere at 3 billion and 2 billion years ago (bya).

Gas	3 bya (%)	2 bya (%)	Present (%)
Nitrogen	91	82	
Oxygen	0	10	
Carbon dioxide	9	8	

(a) Complete the table to show the present day proportion of gases.   **[1]**

(b) Describe how the percentages of the gases in the atmosphere have changed between 3 bya and 2 bya.

Give reasons for the changes.

....

....

....   **[3]**

(c) The activity of humans is believed to have caused the recent increased production of greenhouse gases.

Draw a graph to show the correlation between carbon dioxide emissions and air temperature.

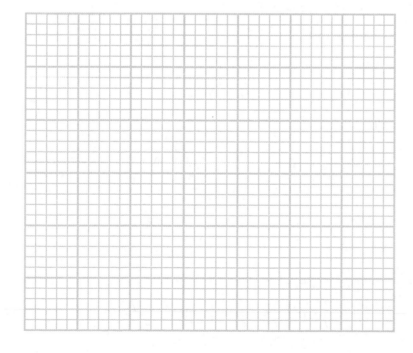

                    [4]

Turn over

18 Look at the flow chart. It shows the steps for the manufacture of sulfuric acid using the contact process.

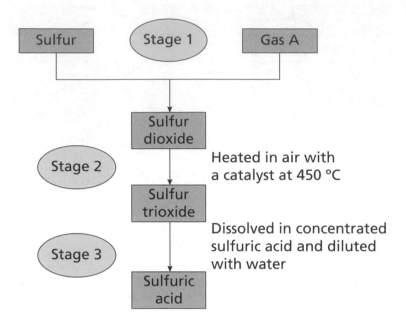

(a) What is the name of Gas A?

.... [1]

(b) Give the **symbol equation** for the reaction taking place in Stage 2.

.... + ... ⇌ .... [3]

(c) Explain the reason for adding a catalyst in Stage 2.

....

.... [2]

19 Look at the following table. It shows the boiling points for six different compounds.

Compound	Number of C atoms	Melting point (°C)	Boiling point (°C)
Water	0	0	100
Methanol	1	−98	65
Ethanol	2	−114	78
Propanol	3	−126	97
Pentanol	5	−109	139
Hexanol	6	−55	

Frasier has a bottle of all the compounds mixed together.

(a) Predict the boiling point of hexanol. ....°C [1]

(b)*Suggest how Frasier can separate the compounds from each other, to get pure samples of each compound.

Explain in detail how your method works.

....

....

....

....

....

....

....

....

....

....

....

....

....

....

.... [6]

Turn over

20 The available land for growing crops worldwide is decreasing.

There is land available, but often it is highly polluted with chemicals.

(a) Phytoextraction is one method of extracting the metals from contaminated soil.

Give **two advantages** and **two disadvantages** of phytoextraction.

advantages

....

.... **[2]**

disadvantages

....

.... **[2]**

(b) Once the soil has been treated, the soil has to be checked for the presence of necessary minerals and ions.

Describe the tests that you would do to test for the presence of iron and chloride ions in the soil and the results you would get.

....

....

....

....

....

.... **[4]**

21 Ethanol (C_2H_5OH) can be produced in a variety of ways.

Cracking 1 mole of C_6H_{14} makes 1 mole of ethene and 1 mole of another hydrocarbon.

(a) Write the complete balanced symbol equation for the cracking of C_6H_{14} into ethene and the other compound

$C_6H_{14} \rightarrow$ .... + ... **[2]**

(b) The ethene produced is then reacted with steam in the presence of a catalyst, to produce 100% ethanol.

Draw the **displayed formula** of the reactants and product for this method of producing ethanol.

$$+ \qquad\qquad\qquad \overset{\text{catalyst}}{\rightleftharpoons}$$

[3]

(c) Look at the graph. It shows the volume of ethanol produced **without** a catalyst.

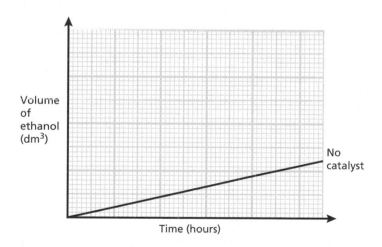

Draw a line on the graph to show the effect of adding a catalyst in this reaction. **[1]**

Turn over

(d) Ethanol can also be produced by the biological method of fermentation.

Yeasts are provided with sugar and water which are converted into ethanol.

The overall reaction can be represented by:

$$C_6H_{12}O_6 \rightarrow 2CH_3CH_2OH + 2CO_2$$

Calculate the relative molecular mass for ethanol.

The relative atomic mass, A_r, of C = 12, O = 16 and H = 1.

[1]

(e) The fermentation process takes time and the ethanol produced is not as pure.

Explain the reasons why chemists may choose the reaction pathway for cracking hydrocarbons to make ethanol, rather than the alternative pathway using fermentation.

[4]

22 Niamh has some soda crystals.

The main chemical in soda crystals is sodium carbonate.

$$Na_2CO_3$$

(a) Calculate the molar mass of 1 mole of sodium carbonate.

The relative atomic mass, A_r, of Na = 23, C = 12 and O = 16

.... [1]

(b) Sodium carbonate reacts with hydrochloric acid to produce sodium chloride, carbon dioxide and water.

Write the **balanced symbol** equation for this reaction.

.... [2]

(c) Describe the tests that Niamh should do to show that soda crystals contain sodium ions and carbonate ions.

(i) Sodium ions test ....

....

....

Result ....

....

(ii) Carbonate ions test ....

....

....

Result ....

.... [4]

23 Ethan is reacting phosphorus trichloride with chlorine.

The symbol equation is shown below:

$$PCl_3(g) + Cl_2(g) \rightleftharpoons PCl_5(g)$$

It is a dynamic equilibrium reaction.

It is exothermic in the direction reactants to products.

(a) Explain what is meant by the term **dynamic equilibrium**.

_____ **[1]**

(b) Ethan wants to maximise the quantity of phosphorus pentachloride produced.

He decides to increase the pressure.

Is Ethan correct? Explain your answer.

_____

_____ **[2]**

(c) Ethan increases the temperature of the system.

Complete the sentences by filling in the missing words.

As the temperature increases the amount of PCl_3 and Cl_2 _____ whilst

the amount of PCl_5 _____. **[1]**

(d) Draw a graph to represent the dynamic equilibrium reached from the forward and reverse reactions using the axes provided.

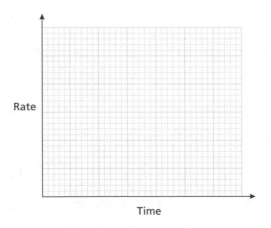

Rate

Time **[2]**

24 Arwen is making calcium carbonate.

Calcium hydroxide reacts with carbon dioxide to form calcium carbonate and water.

The balanced symbol equation for the reaction is:

$$Ca(OH)_2 + CO_2 \rightarrow CaCO_3 + H_2O$$

(a) Calculate the formula mass of $Ca(OH)_2$.

The relative atomic mass, A_r, of Ca = 40.1, O = 16, H = 1 and C = 12.

formula mass = ... [1]

(b) Arwen uses 148 g of $Ca(OH)_2$ in the reaction.

How many moles of $Ca(OH)_2$ is she reacting? Show your working.

number of moles = ... [2]

(c) The amount of $CaCO_3$ actually formed was 103 g.

Calculate the percentage yield.

Show your working. Give your answer to three significant figures.

percentage yield = ... [3]

(d) Arwen needs to make another 200 g of $CaCO_3$.

Given the actual yield in part (c) calculate the mass of $Ca(OH)_2$ she needs to use to make 200 g of $CaCO_3$.

Show your working. Give your answer to three significant figures.

$Ca(OH)_2$ needed = ... [2]

Turn over

Practice Exam Paper 2

25 Magnesium reacts very slowly with cold water to produce an alkaline solution and hydrogen gas.

(a) Write a balanced symbol equation to show the reaction between magnesium and cold water.

.... **[1]**

(b) Describe the test for hydrogen gas.

....

....

.... **[2]**

(c) Magnesium is in Group 2 of the periodic table.

Describe the **difference** in the reaction that would be seen with the element immediately before magnesium in the periodic table.

.... **[1]**

(d) Magnesium cannot be extracted from its ore with carbon.

Explain why this is the case and suggest how magnesium can be extracted from its ore.

....

....

....

....

.... **[3]**

END OF QUESTION PAPER

Answers

Workbook Answers

You are encouraged to show all your working out, as you may be awarded marks for your method even if your final answer is wrong.

Page 148 – Particle Model and Atomic Structure

1. C [1]
2. B [1]
3. a) A = carbon-13 [1]; B = 6 [1]; C = 8 [1]; D = 6 [1]
   b) proton [1]
   c) 60 [1]
4. **Any three from:** there is no indication of the differences in attraction between particles [1]; no difference in size of particles [1]; between the particles is empty space [1]; diagram is only in 2D, not 3D [1]

Page 149 – Purity and Separating Mixtures

1. Each price is for a different purity [1]; the most expensive is the purest grade of glucose / no contaminants [1]
2. D [1]
3. A_r: Mg = 24, O = 16, H = 1, 24 + (2 × 1) + (2 × 16) [1]; = 58 [1]
4. A_r: C = 12, H = 1, O = 16, $\frac{60}{12}$ = 5, $\frac{12}{1}$ = 12, $\frac{16}{16}$ = 1 [1]; $C_5H_{12}O$ [1]
5. Sandy should carry out evaporation / crystallisation [1]; pour the ink into a conical flask / suitable glass container and then gently heat [1]; the liquid component of the ink will evaporate leaving the dried ink pigments [1]

Page 150 – Bonding

1. D [1]
2. A [1]
3. Covalent bonds occur where one electron in the outer shell of each atom [1]; is shared by each of the atoms (two shared electrons = one covalent bond) [1]; ionic bonds occur where the metal element donates an electron or electrons and the non-metal receives the electron(s) [1]; both atoms end up with a full outer shell [1]
4. C [1]
5. A correctly drawn diagram [1]

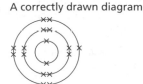

Page 151 – Models of Bonding

1. a) A correctly drawn dot and cross diagram [1]; and displayed formula [1]

Dot and Cross   Displayed Formula

   b) Ball and stick models can give an indication of bond angles / the 3D shape of the molecule. [1]
2. A correctly drawn diagram [1]

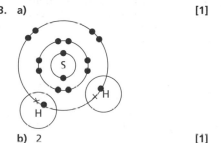

Al^{3+}

3. a) [1]

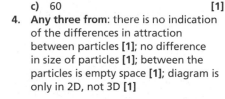

   b) 2 [1]

Page 152 – Properties of Materials

1. C [1]
2. Graphite is made up of repeating carbon atoms that are held together with covalent bonds [1]; it takes a lot of energy to separate carbon atoms covalently joined [1]; water is a covalent molecule made up of only two hydrogens and an oxygen [1]; the water molecules are all joined together by weak forces, making them easy to separate from each other [1]
3. a) Nanoparticles are small particles of a substance in the range of 1–100nm. [1]
   b) Silver nanoparticles have a very large surface area compared to their volume [1]; silver also has antibacterial properties [1]; adding them to the chopping board means that they can kill bacteria present when cutting food [1]
4. A [1]
5. The carbon atoms are each covalently joined to four other carbon atoms [1]; this makes the carbon atoms extremely difficult to separate from each other [1]

Page 153 – Introducing Chemical Reactions

1. a) Be = 1; Al = 2; O = 4 [1]
   b) Al = 2; O = 5; Si = 1 [1]
   c) Ba = 1; P = 2; O = 6 [1]
   d) C = 18; H = 36; O = 2 [1]
2. B [1]
3. $MgCO_3 + 2HNO_3 \rightarrow Mg(NO_3)_2 + CO_2 + H_2O$ [2]
   (1 mark for the correct chemical symbols; 1 mark for correct balancing numbers)

Page 154 – Chemical Equations

1. C [1]
2. The measurement [1]; of the relative amounts of reactants and products in chemical reactions [1]
3. a) $2Br^- \rightarrow Br_2 + 2e^-$ [1]
   b) $H_2 \rightarrow 2H^+ + 2e^-$ [1]
   c) $Fe^{3+} + 3e^- \rightarrow Fe$ [1]
4. a) Spectator ions are ions that are present on both sides of a reaction / the ions do not change / they remain the same. [1]
   b) Na^+ and SO_4^{2-} [1]
   c) $Cu^{2+}(aq) + CO_3^{2-}(aq) \rightarrow CO_3(s)$ [2]
   (1 mark for correct balancing; 1 mark for correct charges)

Page 155 – Moles and Mass

1. B [1]
2. C [1]
3. A_r: Be = 9, Al = 27, O = 16 [1]; formula mass = (1 × 9) + (2 × 27) + (4 × 16) [1]; = 127 [1]
4. Reaction equation: $2H_2(g) + O_2(g) \rightarrow 2H_2O(l)$, 5mol of H_2 will lead to 5mol of H_2O [1]; M_r: H_2O = (2 × 1) + 16 = 18g/mol [1]; 5mol × 18g/mol = 90g [1]

> Look at the equation and work out the ratio of reactant to product. Here it is 2mol of H_2 to 2mol of H_2O, which is a 1:1 ratio. This can then be multiplied to find any amount.

5. $\frac{\text{atomic mass of Rb}}{\text{Avogadro's constant}} = \frac{85g}{6.022 \times 10^{23}}$ [1]; = 1.4×10^{-22}g [1]

Page 156 – Energetics

1. Exothermic reactions release energy to the surroundings [1]; and cause a temperature rise [1]; in this case, the energy given out is used to make light [1]
2. A [1]
3. An endothermic reaction takes in energy from the environment [1]; this means that there will be a temperature drop [1]
4. a) exothermic [1]
   b) endothermic [1]
   c) exothermic [1]
   d) endothermic [1]

Page 157 – Types of Chemical Reactions

1. a) chlorine [1]
    b) oxygen [1]
    c) bromine [1]
    d) oxygen [1]
2. a) iron [1]
    b) carbon [1]
    c) $Fe_2O_3(s) + 3CO(g) \rightarrow$
        $2Fe(s) + 3CO_2(g)$ [2]
        (1 mark for correct reactants; 1 mark for correct products)
3. a) $HCl(aq)$ [1]
    b) $HNO_3(aq)$ [1]
    c) $CH_3CH_2COOH(aq)$ [1]
    d) $HBr(aq)$ [1]
4. B [1]

Page 158 – pH, Acids and Neutralisation

1. Using universal indicator is a qualitative measurement [1]; it is possible for two people to arrive at two different pH levels [1]; a data logger gives a precise, quantitative measurement [1]
2. B [1]; D [1]; E [1]
3. a) C [1]
    b) D [1]
    c) A [1]
    d) B [1]

Page 159 – Electrolysis

1. a) C [1]
    b) Above zinc, but below carbon [1]
2. Metals below carbon can be removed from their ores by the displacement by carbon [1]; more reactive metals will not be displaced [1]; electrolysis separates the molten ions [1]

Page 160 – Predicting Chemical Reactions

1. B [1]
2. C [1]
3. **Any two from:** they can have different charges on their ions [1]; they form compounds with different colours [1]; not as reactive as elements in Groups 1 and 2 [1]
4. Both are in Period 2, which means their outer electrons are close to the nucleus / they have a strong attraction for electrons [1]; oxygen requires 2 more electrons to fill the outer shell, whilst fluorine requires just 1 extra electron [1]; it is easier to attract 1 electron than 2, so fluorine is more reactive [1]

Page 161 – Identifying the Products of Chemical Reactions

1. D [1]
2. Collect the gas in a test tube and place a lit splint into the neck of the test tube [1]; if there is a squeaky pop, hydrogen is present [1]
3. a) copper / Cu^{2+} [1]
    b) sulfate / SO_4^{2-} [1]
    c) $CuSO_4(aq)$ [1]
    d) A reddish brown material forming on the silvery grey nail [1]; and the blue colour of the solution disappearing / fading [1]

Page 162 – Ion Tests and Instrumental Methods of Analysis

1. Add a small amount of nitric acid to acidify the solution [1]; react the solution with a few drops of silver nitrate solution [1]; if a halide (halogen) ion is present a precipitate will form [1]
2. Dip a clean nichrome wire into concentrated hydrochloric acid [1]; dip the wire into the solid that is being tested [1]; and place the wire with the sample into the hottest part of the Bunsen burner flame [1]; note the colour that is produced [1]
3. Cu^{2+} [1]; OH^- [1]
4. B [1]
5. a) Ions are more volatile when converted to chlorides [1]
    b) i) green-blue [1]
        ii) yellow-red [1]
        iii) lilac [1]

Page 163 – Monitoring Chemical Reactions

1. a) C [1]
    b) A [1]; C [1]
    c) Repeat until the results are concordant [1]; and then take the average [1]
2. a) mol/dm^3 is the number of moles of a substance dissolved in $1dm^3$ of solvent [1]
    b) concentration (moles) =
        $$\frac{mass}{relative\ molecular\ mass}\ [1];$$
        $= \frac{109.5}{36.5}$ [1]; = 3 mol/dm^3 [1]

Page 164 – Calculating Yields and Atom Economy

1. a) D [1]
    b) M_r: CuO = 80, amount in moles = $\frac{160}{80}$ = 2 [1]; a 1 : 1 ratio of copper oxide used to copper sulfate formed [1]; M_r: $CuSO_4$ = 160, 2 × 160 = 320g will be formed [1]
    c) Theoretical yield is the yield predicted if all the reactants fully react to make the product [1]
    d) percentage yield =
        $\frac{actual\ yield}{theoretical\ yield} \times 100$ [1]; =
        $\frac{160g}{320g} \times 100$ [1]; = 50% [1]
    e) Desired yield = 640g, efficiency is 50%, therefore, aim for 1280g yield [1]; amount of moles = $\frac{1280g}{160}$ = 8 mol [1]; 8 × 80 = 640g CuO [1]

Page 165 – Controlling Chemical Reactions

1. A [1]
2. C [1]
3. **Any two from:** use a catalyst [1]; increase temperature [1]; increase pressure [1]; increase concentration [1]
4. a) B [1]
    b) The larger the surface area [1]; the greater the opportunity for the reactants to come together and have a successful reaction [1]

Page 166 – Catalysts and Activation Energy

1. a) A steeper line equals a faster rate of reaction. [1]
    b) 45°C [1]
    c) $74cm^3$ [1]
    d) 8min (accept 7.5 min) [1]
    e) Accept any time under 1.5min [1]
2. A catalyst speeds up the rate of reaction [1]; it does this by lowering the activation energy [1]; so the reaction needs less energy to start [1]; it does this without being involved directly / it is not a reactant [1]

Page 167 – Equilibria

1. A reaction that only goes in one direction / cannot go back / cannot re-form reactants [1]
2. C and D [1]
3. C [1]
4. It will have no effect (the number of moles of gas is the same on each side) [1]

> Look at the amount of moles of gas on either side. If they are the same, pressure will not have an effect.

Page 168 – Improving Processes and Products

1. ore [1]
2. C [1]
3. a) Wear eye protection / safety goggles [1]; tilt the tube away from your body [1]
    b) copper(II) oxide + carbon $\rightarrow$
        carbon dioxide + copper [2]
        (1 mark for each correct product)
    c) $2Cu_2O(s) + C(s) \rightarrow$
        $CO_2(g) + 4Cu(s)$ [2]
        (1 mark for correct reactants; 1 mark for correct products)

Page 169 – The Haber Process

1. D [1]
2. hydrogen + nitrogen $\rightleftharpoons$ ammonia [2]
    (1 mark for correct reactants and product; 1 mark for the reversible reaction symbol)
3. Lower temperatures would make the Haber process yield more [1]; but the reaction would be very slow [1]; the pressure could be higher but costs a lot more [1]; a catalyst lowers activation energy, reducing the cost of energy inputs [1]

Page 170 – Life Cycle Assessments, Recycling and Alloys

1. a) Recycling reduces the amount of waste going to landfill [1]; the materials that would be lost are still in circulation [1]; energy costs of dumping the product are reduced [1]

b) A diagram showing two different sizes of atoms in a regular arrangement **[1]**; correctly labelled / with a correct key **[1]**

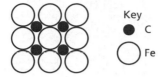

Key
● C
○ Fe

The carbon atoms fit into the gaps between the iron atoms **[1]**; this means that the iron atoms cannot slide over one another **[1]**

Page 171 – Using Materials

1. **a)** C **[1]**
   b) iron + water + oxygen → hydrated iron oxide **[2]** (1 mark for correct reactants; 1 mark for correct product)
   c) Steel corrodes easily **[1]**; therefore it needs a barrier to corrosion **[1]**; e.g. painting, coating in plastic **[1]**

2. Three correctly drawn lines **[2]** (1 mark for one correct line) Poly(ethene) – light and flexible – moulded containers, Polystyrene – light and poor heat conductor – insulation, Polyester – tough and can be coloured – clothing

Page 172 – Organic Chemistry

1. **a)** A **[1]**
   b) C **[1]**
   c) $CH_3CH_2CH_2COOH$ **[2]** (1 mark for three correct hydrocarbon units; 1 mark for correct functional group)
   d) There are double (or triple) bonds between carbon atoms **[1]**
   e) B and D **[1]**
   f) –OH **[1]**
   g) alcohols **[1]**

Page 173 – Organic Compounds and Polymers

1. B **[1]**
2. propene + hydrogen → propane **[1]**
3. C **[1]**

Page 174 – Crude Oil and Fuel Cells

1. A = refinery gases **[1]**
   B = naphtha **[1]**
   C = diesel **[1]**
   D = bitumen **[1]**
2. Short chain hydrocarbons only have a few intermolecular forces keeping them together **[1]**; long chain hydrocarbons have a lot of intermolecular forces keeping the chains together, making it much more difficult to separate them **[1]**
3. Cracking is taking long chain hydrocarbons and breaking them into smaller, more useful short-chained molecules **[1]**
4. **Any two from:** does not produce polluting waste **[1]**; raw ingredients easy to source **[1]**; has a high energy efficiency **[1]**

Page 175 – Interpreting and Interacting with Earth's Systems

1. **a)** The numbers in the table will have been rounded (up and down) **[1]**
   b) D, E, A, B, C **[2]** (1 mark for A after D; 1 mark for C after B)
2. C **[1]**

Page 176 – Air Pollution, Potable Water and Fertilisers

1. Four correctly drawn lines **[3]** (2 marks for two correct lines; 1 mark for one) Carbon monoxide – air poisoning, Sulfur dioxide – kills plants and aquatic life, Nitrogen oxides – photochemical smogs, Particulates – coat surfaces with soot
2. **a)** A = large solids **[1]**; B = small particulates **[1]**
   b) C = bacteria **[1]**; D = chlorine **[1]**
3. Correctly drawn apparatus and process **[1]**; all correctly labelled **[1]**

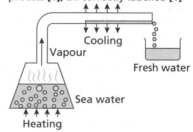

Cooling
Vapour
Fresh water
Sea water
Heating

Page 177 – Practice Exam Paper 1

1. D **[1]**
2. C **[1]**
3. B **[1]**
4. D **[1]**
5. D **[1]**
6. C **[1]**
7. B **[1]**
8. C **[1]**
9. B **[1]**
10. B **[1]**
11. A **[1]**
12. D **[1]**
13. C **[1]**
14. D **[1]**
15. A **[1]**
16. **(a)** The boiling point tends to increase as you go across the period from Groups 1 to 4 **[1]**; with a drop in Groups 5, 6, 7 and 0 **[1]**
    (b) **Any two from:** the period indicates the number of electron shells around the nucleus **[1]**; as you move across a period, the elements on the left-hand side are more likely to be metals, as they donate electrons **[1]**; those on the right are non-metals, as they receive electrons **[1]**; there is a change from metallic bonding to covalent bonding **[1]**; as you move right across the period in the metals section, the metals become less reactive relative to each other **[1]**; as you move left in the non-metals section, the non-metals become less reactive relative to each other **[1]**
    (c) Particle X is an atom, Y is an ion **[1]**; the atomic number indicates the number of protons. If the number of electrons, which can be found from the electronic structure, matches the atomic number it must be neutral, i.e. an atom **[1]** (Or reverse argument)
    (d) (i) The atom is not to scale **[1]**; the proton is drawn too large **[1]**; and the electron is too close to the nucleus **[1]**
    (ii) One early model had the charges distributed evenly throughout the atom **[1]**; Bohr had electrons in discrete shells / energy levels **[1]**

17. **(a)** Graphene is a single lattice of atoms, making it very thin **[1]**; it also allows electrons to travel through the connected atoms / it conducts electricity **[1]**
    (b) Drawing showing at least two layers of carbon atoms (arranged with three bonds per carbon atom) **[1]**; labelled intermolecular / weak forces between layers **[1]**

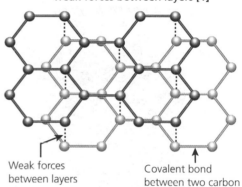

Weak forces between layers

Covalent bond between two carbon atoms within a layer

    (c) We did not have the instrumental analysis / ability to detect the new forms **[1]**; there is far more diamond and graphite available today **[1]**

18. **(a)** **Any two pairs from:** the reaction is endothermic **[1]**; as the reactants have less energy than the products **[1]** OR A is the activation energy **[1]**; which is the energy supplied to start the reaction **[1]** OR B is the overall energy change for the reaction **[1]**; the value of B is positive **[1]**
    (b) (i) broken = 3, made = 4 **[1]**
    (ii) When bonds are broken: endothermic / energy is taken in **[1]**; when bonds are made: exothermic / energy is released **[1]**
    (iii) energy needed to break bonds = +1358kJ/mol **[1]**; energy released when new bonds form = –1856kJ/mol **[1]**; energy change = –498kJ/mol / 498kJ/mol given out **[1]**

19. **(a)** The copper ions have a 2+ charge **[1]**; they will move to the negative electrode / cathode where they will pick up 2 electrons **[1]**; this means that the outer shell will have 2 electrons in it / it is the same as the elemental form of copper **[1]**
    (b) $Cu^{2+}(aq) + 2e^-$ **[1]**; $\rightarrow Cu(s)$ **[1]**

(c) Bubbles of oxygen may be seen [1]
(d) Replace the solution with silver nitrate [1]; replace the cathode with the earring [1]
20. (a) **Any two pairs from:** he should dip the tip into concentrated hydrochloric acid [1]; the resulting metal chloride is more volatile [1] OR Rather than dipping into the solution, he should dip directly into the solid salt [1]; this will increase the amount that will be added to the flame making it easier to see the result [1] OR He should ensure that the Bunsen flame is on the blue flame setting because this is the hottest flame [1]; and the colour of a blue flame will not interfere with the colour the metal ion emits [1]
(b) lithium / Li^+ [1]
(c) (i) bromide / Br^- [1]
(ii) $Ag^+(aq) + Br^-(aq) \rightarrow AgBr(s)$ [1]
(d) Diagram shows equal distribution of Ag^+ and Br^- ions in solid arrangement [1]; key / indication of which atom is which [1]

● Ag⁺
● Br⁻

21. (a) 1–100nm [1]
(b) Nanoparticles have a very large surface area [1]; so the bacteria produced from food spoilage are more likely to come into contact with them [1]
(c) **Any three from:** as the nanoparticles are embedded into food packaging they are more likely to enter our body accidentally [1]; nanoparticles can act as catalysts [1]; we don't know what the effect will be in the body [1]; our immune systems may not cope with the particles [1]; cannot be removed from the body once they are there / undetectable [1]
22. (a)***Answer to also include discussion of the dissociation constant, for example:** the dissociation constant is likely to be related to how easily an acid dissociates into ions [1]; there are weak and strong acids and those that are strong dissociate easily (e.g. H_2SO_4) [1]; so strong acids have a larger dissociation constant [1]; (accept alternative comments about weak acids)

Answer to also include discussion of the experiment, for example: the strengths of the acids could be compared by titration with an alkali / by adding a volume of alkali to a known volume of acid gradually until the acid is neutralised [1]; the stronger the acid, the greater the volume of alkali needed to neutralise it [1]; this is repeated until concordant results are achieved [1]

Some exam questions have an asterisk (*) alongside the number. This means that marks are going to be awarded based on the level of response that you give. The examiner is looking for a factual explanation using scientific words. To gain full marks, your answer must be written in a logical and coherent way, with all the key points clearly explained, supported and developed.

(b) 10 times stronger [1]
23. (a) (i) 18.4g [1]
(ii) number of moles of magnesium = $\frac{2.4g}{24g/mol}$ = 0.1 [1]; number of moles of X = 0.1mol [1]; number of moles of magnesium X = 0.1mol [1]
(b) $Mg + X_2 \rightarrow MgX_2$ [2] (1 mark for correct reactants; 1 mark for correct products)
(c) Mg^{2+} and Br^- (allow X^-) [1]
24. (a) $2Al(s) + Fe_2O_3(s) \rightarrow 2Fe(s) + Al_2O_3(s)$ [1]
(b) **Any two from:** eye protection [1]; carry out behind a safety screen [1]; use a fuse to start the reaction [1]; do not do the reaction inside [1]
(c) oxidised = Al [1]; reduced = Fe [1]
(d) Formula mass of Al_2O_3 = 102, therefore 10.2g is 0.1 moles [1]; ratio of Fe_2O_3 to Al_2O_3 = 1 : 1 so 0.1 moles of $Fe_2O_3(s)$ [1]; formula mass of $Fe_2O_3(s)$ = 159.6, therefore 16.0g needed (to the nearest gram) [1]
25. (a) $\frac{805}{6.02 \times 10^{23}}$ = 1.34×10^{-21}g [1]
(b) $\frac{805}{342}$ = 2.35 times heavier [1]

Page 197 – Practice Exam Paper 2

1. B [1]
2. D [1]
3. D [1]
4. B [1]
5. D [1]
6. B [1]
7. A [1]
8. B [1]
9. D [1]
10. A [1]
11. A [1]
12. C [1]
13. C [1]
14. B [1]
15. D [1]

16. (a) 2.8.8.1 [1]
(b) Diagram that shows positive ions arranged in solid-like formation [1]; indication of sea of delocalised electrons [1]

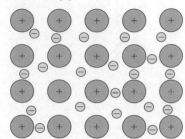

(c) The potassium moves around rapidly on the surface of the water [1]; and a gas ignites with a lilac flame [1]
(d) potassium + water → potassium hydroxide + hydrogen [2] (1 mark for each correct product)
17. (a) N_2 = 78, O_2 = 21, CO_2 = 0.04 (accept 0.03) [1]
(b) The amount of nitrogen has dropped by 9%, because of the increase in oxygen levels [1]; there was no oxygen 3bya but 2bya 10% of the atmosphere was oxygen, due to the appearance of photosynthesising plants [1]; the amount of carbon dioxide remained relatively constant, this was due to volcanic activity [1]
(c) A graph with x-axis labelled time and y-axis labelled Amount of CO_2 and Air Temperature (may be on either side) [1]; labelled lines for air temp and CO_2 [1]; shape resembles a hockey stick [1]; timing indicates CO_2 levels increased from industrial era / 1800s [1]

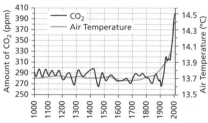

18. (a) Oxygen [1]
(b) $2SO_2$ [1] + O_2 [1] ⇌ $2SO_3$ [1]
(c) The reaction is slow [1]; the catalyst lowers the activation energy [1]
19. (a) Accept answers above 190°C [1]
(b)*The compounds in the table are all alcohols (except water), and each alcohol contains one to six carbon atoms – the longer the chain length, the higher the melting point / boiling point [1]; the alcohols can be separated via (fractional) distillation [1]; this involves heating the mixture until the boiling point of the shortest chained alcohol is reached and then cooling the vapour to liquid so that it can be collected

(this is repeated up to propanol) **[1]**; pentanol and hexanol can be extracted using gas chromatography **[1]**; this involves injecting a sample of the alcohols into a solid medium that can separate gaseous compounds **[1]**; once collected, the purity of each alcohol can be tested by measuring the boiling points and checking against published data **[1]**

20. **(a)** Advantages **(any two from)**: low cost **[1]**; no maintenance **[1]**; wildlife can flourish at the site **[1]**; aesthetically pleasing **[1]** Disadvantages **(any two from)**: takes a long time **[1]**; limited to contamination that can be accessed by the plant roots **[1]**; burning / disposing of the plants could release contamination **[1]**; there is a maximum concentration of contaminant the plants can handle **[1]**

(b) The test for iron ions is to add sodium hydroxide solution **[1]**; if the precipitate is a green colour, it is iron(II) and if it goes a red colour it is iron(III) **[1]**; the test for chloride ions is to add dilute nitric acid and silver nitrate solution **[1]**; if a white precipitate is formed, the chloride ion is present **[1]**

21. **(a)** C_2H_4 **[1]**; C_4H_{10} **[1]** (either order)

(b)
H   H
 \ /
  C=C   H H
 / \   +   \ /
H   H   O

⇌   H—C—C—O—H (ethanol structure with H H / H H)

[3] (1 mark for each correct reactant; 1 mark for correct product)

(c) **[1]**

Volume of ethanol (dm³) vs Time (hours) — curves labelled "Catalyst" and "No catalyst"

(d) 46 **[1]**

(e) **Any four from:** cracking is faster **[1]**; alcohol can be continuously produced using cracking **[1]**; high purity alcohol is produced by cracking **[1]**; fermenting produces lots of other products **[1]**; alcohol can only be produced in batches by fermentation **[1]**; can't be as pure, as yeast used in fermentation will die if too much alcohol **[1]**

22. **(a)** 106 g/mol **[1]**

(b) $Na_2CO_3 + 2HCl \rightarrow$ $2NaCl + H_2O + CO_2$ **[2]** (1 mark for correct formulae; 1 mark for correct balancing)

(c) **(i)** Perform a flame test **[1]**; Yellow/Orange flame observed **[1]**

(ii) Test: add hydrochloric acid – bubbles are formed. Test the gas given off for carbon dioxide by passing it into limewater **[1]**; Result: carbon dioxide turns limewater milky **[1]**

23. **(a)** The rate of forward reaction equals the rate of the reverse reaction **[1]**

(b) Yes - increasing the pressure shifts the reaction to the side with the fewer molecules **[1]**; in this case there are two molecules on the reactant side and only one on the product side **[1]**

(c) Increases, decreases **[1]**

(d) Two curves drawn leading to equilibrium **[1]**; labelled correctly **[1]**

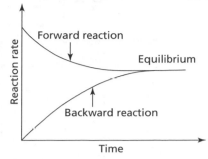

Reaction rate vs Time graph — curves labelled "Forward reaction", "Backward reaction", "Equilibrium"

24. **(a)** 74.1g/mol **[1]**

(b) $\dfrac{148}{74.1}$ **[1]**; = 2 **[1]**

(c) Theoretical yield of $CaCO_3$ = 2 × (40.1 + 12 + (16 × 3)) = 200.2 **[1]**; percentage yield = $\dfrac{\text{actual yield}}{\text{theoretical yield}} \times 100$ **[1]**; = $\dfrac{103}{200.2} \times 100 = 51.4\%$ **[1]**

> 2 moles of $Ca(OH)_2$ are used, so 2 moles of $CaCO_3$ are expected, i.e. theoretical yield = 2 × molar mass of $CaCO_3$

(d) If 103g of $CaCO_3$ is made from 148g of $Ca(OH)_2$, then 1g will be made from $\dfrac{148}{103g}$ **[1]**; and 200g will be made from $\dfrac{148}{103g} \times 200g = 287g$ **[1]**

25. **(a)** $Mg(s) + 2H_2O(l) \rightarrow$ $Mg(OH)_2(aq) + H_2(g)$ **[1]**

(b) Insert a lit splint into the top of a test tube **[1]**; if it makes a squeaky pop, hydrogen is present **[1]**

(c) The reaction would be faster / more vigorous **[1]**

(d) Magnesium is higher than carbon in the reactivity series **[1]**; electrolysis **[1]**; of the molten ore is required **[1]**

The Periodic Table

(1)	(2)											(3)	(4)	(5)	(6)	(7)	(0)
1	2	3	4	5	6	7	8	9	10	11	12	13	14	15	16	17	18
1 **H** hydrogen 1.0																	2 **He** helium 4.0
3 **Li** lithium 6.9	4 **Be** beryllium 9.0											5 **B** boron 10.8	6 **C** carbon 12.0	7 **N** nitrogen 14.0	8 **O** oxygen 16.0	9 **F** fluorine 19.0	10 **Ne** neon 20.2
11 **Na** sodium 23.0	12 **Mg** magnesium 24.3											13 **Al** aluminum 27.0	14 **Si** silicon 28.1	15 **P** phosphorus 31.0	16 **S** sulfur 32.1	17 **Cl** chlorine 35.5	18 **Ar** argon 39.9
19 **K** potassium 39.1	20 **Ca** calcium 40.1	21 **Sc** scandium 45.0	22 **Ti** titanium 47.9	23 **V** vanadium 50.9	24 **Cr** chromium 52.0	25 **Mn** manganese 54.9	26 **Fe** iron 55.8	27 **Co** cobalt 58.9	28 **Ni** nickel 58.7	29 **Cu** copper 63.5	30 **Zn** zinc 65.4	31 **Ga** gallium 69.7	32 **Ge** germanium 72.6	33 **As** arsenic 74.9	34 **Se** selenium 79.0	35 **Br** bromine 79.9	36 **Kr** krypton 83.8
37 **Rb** rubidium 85.5	38 **Sr** strontium 87.6	39 **Y** yttrium 88.9	40 **Zr** zirconium 91.2	41 **Nb** niobium 92.9	42 **Mo** molybdenum 95.9	43 **Tc** technetium	44 **Ru** ruthenium 101.1	45 **Rh** rhodium 102.9	46 **Pd** palladium 106.4	47 **Ag** silver 107.9	48 **Cd** cadmium 112.4	49 **In** indium 114.8	50 **Sn** tin 118.7	51 **Sb** antimony 121.8	52 **Te** tellurium 127.6	53 **I** iodine 126.9	54 **Xe** xenon 131.3
55 **Cs** cesium 132.9	56 **Ba** barium 137.3	57-71 lanthanides	72 **Hf** hafnium 178.5	73 **Ta** tantalum 180.9	74 **W** tungsten 183.8	75 **Re** rhenium 186.2	76 **Os** osmium 190.2	77 **Ir** iridium 192.2	78 **Pt** platinum 195.1	79 **Au** gold 197.0	80 **Hg** mercury 200.5	81 **Tl** thallium 204.4	82 **Pb** lead 207.2	83 **Bi** bismuth 209.0	84 **Po** polonium	85 **At** astatine	86 **Rn** radon
87 **Fr** francium	88 **Ra** radium	89-103 actinides	104 **Rf** rutherfordium	105 **Db** dubnium	106 **Sg** seaborgium	107 **Bh** bohrium	108 **Hs** hassium	109 **Mt** meitnerium	110 **Ds** darmstadtium	111 **Rg** roentgenium	112 **Cn** copernicium		114 **Fl** flerovium		116 **Lv** livermorium		

Key

atomic number	
symbol	
name	
relative atomic mass	